Lecture Notes in Mathematics

Edited by A. Dold and B. Eckmann

1381

Jan-Olov Strömberg
Alberto Torchinsky

Weighted Hardy Spaces

Springer-Verlag
Berlin Heidelberg New York London Paris Tokyo Hong Kong

Authors

Jan-Olov Strömberg
University of Tromsø, Institute of Mathematical and Physical Sciences
9001 Tromsø, Norway

Alberto Torchinsky
Indiana University, Department of Mathematics
Bloomington, IN 47405, USA

Mathematics Subject Classification (1980): 42 B 30

ISBN 3-540-51402-3 Springer-Verlag Berlin Heidelberg New York
ISBN 0-387-51402-3 Springer-Verlag New York Berlin Heidelberg

© Springer-Verlag Berlin Heidelberg 1989
Printed in Germany

Printing and binding: Druckhaus Beltz, Hemsbach/Bergstr.
2146/3140-543210 – Printed on acid-free paper

Vol. 1232: P.C. Schuur, Asymptotic Analysis of Soliton Problems. VIII, 180 pages. 1986.

Vol. 1233: Stability Problems for Stochastic Models. Proceedings, 1985. Edited by V.V. Kalashnikov, B. Penkov and V.M. Zolotarev. VI, 223 pages. 1986.

Vol. 1234: Combinatoire énumérative. Proceedings, 1985. Edité par G. Labelle et P. Leroux. XIV, 387 pages. 1986.

Vol. 1235: Séminaire de Théorie du Potentiel, Paris, No. 8. Directeurs: M. Brelot, G. Choquet et J. Deny. Rédacteurs: F. Hirsch et G. Mokobodzki. III, 209 pages. 1987.

Vol. 1236: Stochastic Partial Differential Equations and Applications. Proceedings, 1985. Edited by G. Da Prato and L. Tubaro. V, 257 pages. 1987.

Vol. 1237: Rational Approximation and its Applications in Mathematics and Physics. Proceedings, 1985. Edited by J. Gilewicz, M. Pindor and W. Siemaszko. XII, 350 pages. 1987.

Vol. 1238: M. Holz, K.-P. Podewski and K. Steffens, Injective Choice Functions. VI, 183 pages. 1987.

Vol. 1239: P. Vojta, Diophantine Approximations and Value Distribution Theory. X, 132 pages. 1987.

Vol. 1240: Number Theory, New York 1984–85. Seminar. Edited by D.V. Chudnovsky, G.V. Chudnovsky, H. Cohn and M.B. Nathanson. V, 324 pages. 1987.

Vol. 1241: L. Gårding, Singularities in Linear Wave Propagation. III, 125 pages. 1987.

Vol. 1242: Functional Analysis II, with Contributions by J. Hoffmann-Jørgensen et al. Edited by S. Kurepa, H. Kraljević and D. Butković. VII, 432 pages. 1987.

Vol. 1243: Non Commutative Harmonic Analysis and Lie Groups. Proceedings, 1985. Edited by J. Carmona, P. Delorme and M. Vergne. V, 309 pages. 1987.

Vol. 1244: W. Müller, Manifolds with Cusps of Rank One. XI, 158 pages. 1987.

Vol. 1245: S. Rallis, L-Functions and the Oscillator Representation. XVI, 239 pages. 1987.

Vol. 1246: Hodge Theory. Proceedings, 1985. Edited by E. Cattani, F. Guillén, A. Kaplan and F. Puerta. VII, 175 pages. 1987.

Vol. 1247: Séminaire de Probabilités XXI. Proceedings. Edité par J. Azéma, P.A. Meyer et M. Yor. IV, 579 pages. 1987.

Vol. 1248: Nonlinear Semigroups, Partial Differential Equations and Attractors. Proceedings, 1985. Edited by T.L. Gill and W.W. Zachary. IX, 185 pages. 1987.

Vol. 1249: I. van den Berg, Nonstandard Asymptotic Analysis. IX, 187 pages. 1987.

Vol. 1250: Stochastic Processes – Mathematics and Physics II. Proceedings 1985. Edited by S. Albeverio, Ph. Blanchard and L. Streit. VI, 359 pages. 1987.

Vol. 1251: Differential Geometric Methods in Mathematical Physics. Proceedings, 1985. Edited by P.L. García and A. Pérez-Rendón. VII, 300 pages. 1987.

Vol. 1252: T. Kaise, Représentations de Weil et GL$_2$ Algèbres de division et GL$_n$. VII, 203 pages. 1987.

Vol. 1253: J. Fischer, An Approach to the Selberg Trace Formula via the Selberg Zeta-Function. III, 184 pages. 1987.

Vol. 1254: S. Gelbart, I. Piatetski-Shapiro, S. Rallis. Explicit Constructions of Automorphic L-Functions. VI, 152 pages. 1987.

Vol. 1255: Differential Geometry and Differential Equations. Proceedings, 1985. Edited by C. Gu, M. Berger and R.L. Bryant. XII, 243 pages. 1987.

Vol. 1256: Pseudo-Differential Operators. Proceedings, 1986. Edited by H.O. Cordes, B. Gramsch and H. Widom. X, 479 pages. 1987.

Vol. 1257: X. Wang, On the C*-Algebras of Foliations in the Plane. V, 165 pages. 1987.

Vol. 1258: J. Weidmann, Spectral Theory of Ordinary Differential Operators. VI, 303 pages. 1987.

Vol. 1259: F. Cano Torres, Desingularization Strategies for Three-Dimensional Vector Fields. IX, 189 pages. 1987.

Vol. 1260: N.H. Pavel, Nonlinear Evolution Operators and Semigroups. VI, 285 pages. 1987.

Vol. 1261: H. Abels, Finite Presentability of S-Arithmetic Groups. Compact Presentability of Solvable Groups. VI, 178 pages. 1987.

Vol. 1262: E. Hlawka (Hrsg.), Zahlentheoretische Analysis II. Seminar, 1984–86. V, 158 Seiten. 1987.

Vol. 1263: V.L. Hansen (Ed.), Differential Geometry. Proceedings, 1985. XI, 288 pages. 1987.

Vol. 1264: Wu Wen-tsün, Rational Homotopy Type. VIII, 219 pages. 1987.

Vol. 1265: W. Van Assche, Asymptotics for Orthogonal Polynomials. VI, 201 pages. 1987.

Vol. 1266: F. Ghione, C. Peskine, E. Sernesi (Eds.), Space Curves. Proceedings, 1985. VI, 272 pages. 1987.

Vol. 1267: J. Lindenstrauss, V.D. Milman (Eds.), Geometrical Aspects of Functional Analysis. Seminar. VII, 212 pages. 1987.

Vol. 1268: S.G. Krantz (Ed.), Complex Analysis. Seminar, 1986. VII, 195 pages. 1987.

Vol. 1269: M. Shiota, Nash Manifolds. VI, 223 pages. 1987.

Vol. 1270: C. Carasso, P.-A. Raviart, D. Serre (Eds.), Nonlinear Hyperbolic Problems. Proceedings, 1986. XV, 341 pages. 1987.

Vol. 1271: A.M. Cohen, W.H. Hesselink, W.L.J. van der Kallen, J.R. Strooker (Eds.), Algebraic Groups Utrecht 1986. Proceedings. XII, 284 pages. 1987.

Vol. 1272: M.S. Livšic, L.L. Waksman, Commuting Nonselfadjoint Operators in Hilbert Space. III, 115 pages. 1987.

Vol. 1273: G.-M. Greuel, G. Trautmann (Eds.), Singularities, Representation of Algebras, and Vector Bundles. Proceedings, 1985. XIV. 383 pages. 1987.

Vol. 1274: N. C. Phillips, Equivariant K-Theory and Freeness of Group Actions on C*-Algebras. VIII, 371 pages. 1987.

Vol. 1275: C.A. Berenstein (Ed.), Complex Analysis I. Proceedings, 1985–86. XV, 331 pages. 1987.

Vol. 1276: C.A. Berenstein (Ed.), Complex Analysis II. Proceedings, 1985–86. IX, 320 pages. 1987.

Vol. 1277: C.A. Berenstein (Ed.), Complex Analysis III. Proceedings, 1985–86. X, 350 pages. 1987.

Vol. 1278: S.S. Koh (Ed.), Invariant Theory. Proceedings, 1985. V, 102 pages. 1987.

Vol. 1279: D. Ieşan, Saint-Venant's Problem. VIII, 162 Seiten. 1987.

Vol. 1280: E. Neher, Jordan Triple Systems by the Grid Approach. XII, 193 pages. 1987.

Vol. 1281: O.H. Kegel, F. Menegazzo, G. Zacher (Eds.), Group Theory. Proceedings, 1986. VII, 179 pages. 1987.

Vol. 1282: D.E. Handelman, Positive Polynomials, Convex Integral Polytopes, and a Random Walk Problem. XI, 136 pages. 1987.

Vol. 1283: S. Mardešić, J. Segal (Eds.), Geometric Topology and Shape Theory. Proceedings, 1986. V, 261 pages. 1987.

Vol. 1284: B.H. Matzat, Konstruktive Galoistheorie. X, 286 pages. 1987.

Vol. 1285: I.W. Knowles, Y. Saitō (Eds.), Differential Equations and Mathematical Physics. Proceedings, 1986. XVI, 499 pages. 1987.

Vol. 1286: H.R. Miller, D.C. Ravenel (Eds.), Algebraic Topology. Proceedings, 1986. VII, 341 pages. 1987.

Vol. 1287: E.B. Saff (Ed.), Approximation Theory, Tampa. Proceedings, 1985–1986. V, 228 pages. 1987.

Vol. 1288: Yu. L. Rodin, Generalized Analytic Functions on Riemann Surfaces. V, 128 pages, 1987.

Vol. 1289: Yu. I. Manin (Ed.), K-Theory, Arithmetic and Geometry. Seminar, 1984–1986. V, 399 pages. 1987.

Vol. 1290: G. Wüstholz (Ed.), Diophantine Approximation and Transcendence Theory. Seminar, 1985. V, 243 pages. 1987.

Vol. 1291: C. Mœglin, M.-F. Vignéras, J.-L. Waldspurger, Correspondances de Howe sur un Corps p-adique. VII, 163 pages. 1987

Vol. 1292: J.T. Baldwin (Ed.), Classification Theory. Proceedings, 1985. VI, 500 pages. 1987.

Vol. 1293: W. Ebeling, The Monodromy Groups of Isolated Singularities of Complete Intersections. XIV, 153 pages. 1987.

Vol. 1294: M. Queffélec, Substitution Dynamical Systems – Spectral Analysis. XIII, 240 pages. 1987.

Vol. 1295: P. Lelong, P. Dolbeault, H. Skoda (Réd.), Séminaire d'Analyse P. Lelong – P. Dolbeault – H. Skoda. Seminar, 1985/1986. VII, 283 pages. 1987.

Vol. 1296: M.-P. Malliavin (Ed.), Séminaire d'Algèbre Paul Dubreil et Marie-Paule Malliavin. Proceedings, 1986. IV, 324 pages. 1987.

Vol. 1297: Zhu Y.-l., Guo B.-y. (Eds.), Numerical Methods for Partial Differential Equations. Proceedings, XI, 244 pages. 1987.

Vol. 1298: J. Aguadé, R. Kane (Eds.), Algebraic Topology, Barcelona 1986. Proceedings, X, 255 pages. 1987.

Vol. 1299: S. Watanabe, Yu.V. Prokhorov (Eds.), Probability Theory and Mathematical Statistics. Proceedings, 1986. VIII, 589 pages. 1988.

Vol. 1300: G.B. Seligman, Constructions of Lie Algebras and their Modules. VI, 190 pages. 1988.

Vol. 1301: N. Schappacher, Periods of Hecke Characters. XV, 160 pages. 1988.

Vol. 1302: M. Cwikel, J. Peetre, Y. Sagher, H. Wallin (Eds.), Function Spaces and Applications. Proceedings, 1986. VI, 445 pages. 1988.

Vol. 1303: L. Accardi, W. von Waldenfels (Eds.), Quantum Probability and Applications III. Proceedings, 1987. VI, 373 pages. 1988.

Vol. 1304: F.Q. Gouvêa, Arithmetic of p-adic Modular Forms. VIII, 121 pages. 1988.

Vol. 1305: D.S. Lubinsky, E.B. Saff, Strong Asymptotics for Extremal Polynomials Associated with Weights on ℝ. VII, 153 pages. 1988.

Vol. 1306: S.S. Chern (Ed.), Partial Differential Equations. Proceedings, 1986. VI, 294 pages. 1988.

Vol. 1307: T. Murai, A Real Variable Method for the Cauchy Transform, and Analytic Capacity. VIII, 133 pages. 1988.

Vol. 1308: P. Imkeller, Two-Parameter Martingales and Their Quadratic Variation. IV, 177 pages. 1988.

Vol. 1309: B. Fiedler, Global Bifurcation of Periodic Solutions with Symmetry. VIII, 144 pages. 1988.

Vol. 1310: O.A. Laudal, G. Pfister, Local Moduli and Singularities. V, 117 pages. 1988.

Vol. 1311: A. Holme, R. Speiser (Eds.), Algebraic Geometry, Sundance 1986. Proceedings, VI, 320 pages. 1988.

Vol. 1312: N.A. Shirokov, Analytic Functions Smooth up to the Boundary. III, 213 pages. 1988.

Vol. 1313: F. Colonius, Optimal Periodic Control. VI, 177 pages. 1988.

Vol. 1314: A. Futaki, Kähler-Einstein Metrics and Integral Invariants. IV, 140 pages. 1988.

Vol. 1315: R.A. McCoy, I. Ntantu, Topological Properties of Spaces of Continuous Functions. IV, 124 pages. 1988.

Vol. 1316: H. Korezlioglu, A.S. Ustunel (Eds.), Stochastic Analysis and Related Topics. Proceedings, 1986. V, 371 pages. 1988.

Vol. 1317: J. Lindenstrauss, V.D. Milman (Eds.), Geometric Aspects of Functional Analysis. Seminar, 1986–87. VII, 289 pages. 1988.

Vol. 1318: Y. Felix (Ed.), Algebraic Topology – Rational Homotopy. Proceedings, 1986. VIII, 245 pages. 1988

Vol. 1319: M. Vuorinen, Conformal Geometry and Quasiregular Mappings. XIX, 209 pages. 1988.

Vol. 1320: H. Jürgensen, G. Lallement, H.J. Weinert (Eds.), Se groups, Theory and Applications. Proceedings, 1986. X, 416 pag 1988.

Vol. 1321: J. Azéma, P.A. Meyer, M. Yor (Eds.), Séminaire Probabilités XXII. Proceedings. IV, 600 pages. 1988.

Vol. 1322: M. Métivier, S. Watanabe (Eds.), Stochastic Analy Proceedings, 1987. VII, 197 pages. 1988.

Vol. 1323: D.R. Anderson, H.J. Munkholm, Boundedly Contro Topology. XII, 309 pages. 1988.

Vol. 1324: F. Cardoso, D.G. de Figueiredo, R. Iório, O. Lopes (E Partial Differential Equations. Proceedings, 1986. VIII, 433 pa 1988.

Vol. 1325: A. Truman, I.M. Davies (Eds.), Stochastic Mechanics Stochastic Processes. Proceedings, 1986. V, 220 pages. 1988.

Vol. 1326: P.S. Landweber (Ed.), Elliptic Curves and Modular Form Algebraic Topology. Proceedings, 1986. V, 224 pages. 1988.

Vol. 1327: W. Bruns, U. Vetter, Determinantal Rings. VII,236 pa 1988.

Vol. 1328: J.L. Bueso, P. Jara, B. Torrecillas (Eds.), Ring The Proceedings, 1986. IX, 331 pages. 1988.

Vol. 1329: M. Alfaro, J.S. Dehesa, F.J. Marcellan, J.L. Rubio Francia, J. Vinuesa (Eds.): Orthogonal Polynomials and their App tions. Proceedings, 1986. XV, 334 pages. 1988.

Vol. 1330: A. Ambrosetti, F. Gori, R. Lucchetti (Eds.), Mathema Economics. Montecatini Terme 1986. Seminar. VII, 137 pages. 19

Vol. 1331: R. Bamón, R. Labarca, J. Palis Jr. (Eds.), Dynam Systems, Valparaiso 1986. Proceedings. VI, 250 pages. 1988.

Vol. 1332: E. Odell, H. Rosenthal (Eds.), Functional Analysis. ceedings, 1986–87. V, 202 pages. 1988.

Vol. 1333: A.S. Kechris, D.A. Martin, J.R. Steel (Eds.), Cabal Sem 81–85. Proceedings, 1981–85. V, 224 pages. 1988.

Vol. 1334: Yu.G. Borisovich, Yu. E. Gliklikh (Eds.), Global Anal – Studies and Applications III. V, 331 pages. 1988.

Vol. 1335: F. Guillén, V. Navarro Aznar, P. Pascual-Gainza, F. Pue Hyperrésolutions cubiques et descente cohomologique. XII, pages. 1988.

Vol. 1336: B. Helffer, Semi-Classical Analysis for the Schrödin Operator and Applications. V, 107 pages. 1988.

Vol. 1337: E. Sernesi (Ed.), Theory of Moduli. Seminar, 1985. VIII, pages. 1988.

Vol. 1338: A.B. Mingarelli, S.G. Halvorsen, Non-Oscillation Dom of Differential Equations with Two Parameters. XI, 109 pages. 1988

Vol. 1339: T. Sunada (Ed.), Geometry and Analysis of Manifol Procedings, 1987. IX, 277 pages. 1988.

Vol. 1340: S. Hildebrandt, D.S. Kinderlehrer, M. Miranda (Ed Calculus of Variations and Partial Differential Equations. Proceedin 1986. IX, 301 pages. 1988.

Vol. 1341: M. Dauge, Elliptic Boundary Value Problems on Cor Domains. VIII, 259 pages. 1988.

Vol. 1342: J.C. Alexander (Ed.), Dynamical Systems. Proceedin 1986–87. VIII, 726 pages. 1988.

Vol. 1343: H. Ulrich, Fixed Point Theory of Parametrized Equivar Maps. VII, 147 pages. 1988.

Vol. 1344: J. Král, J. Lukeš, J. Netuka, J. Veselý (Eds.), Poten Theory – Surveys and Problems. Proceedings, 1987. VIII, 271 pag 1988.

Vol. 1345: X. Gomez-Mont, J. Seade, A. Verjovski (Eds.), Holomorp Dynamics. Proceedings, 1986. VII, 321 pages. 1988.

Vol. 1346: O. Ya. Viro (Ed.), Topology and Geometry – Ro Seminar. XI, 581 pages. 1988.

Vol. 1347: C. Preston, Iterates of Piecewise Monotone Mappings an Interval. V, 166 pages. 1988.

Vol. 1348: F. Borceux (Ed.), Categorical Algebra and its Applicatio Proceedings, 1987. VIII, 375 pages. 1988.

Vol. 1349: E. Novak, Deterministic and Stochastic Error Bound Numerical Analysis. V, 113 pages. 1988.

Preface

A considerable development of harmonic analysis in the last few years has been centered around a function space shown in a new light, the functions of bounded mean oscillation, and the weighted inequalities for classical operators. The new techniques introduced by C. Fefferman and E. Stein and B. Muckenhoupt are basic in these areas; for further details the reader may consult the monographs of García-Cuerva and Rubio de Francia [1985] and Torchinsky [1986]. It is our purpose here to further develop some of these results in the general setting of the weighted Hardy spaces, and to discuss some applications. The origin of these notes is the announcement in Strömberg and Torchinsky [1980], and the course given by the first author at Rutgers University in the academic year 1985-1986.

A word about the content of the notes. In Chapter I we introduce the notion of weighted measures in the general context of homogenous spaces; the results discussed here include the theory of A_p weights. Chapter II deals with the Jones decomposition of these weights including a novel feature, namely, the control of the doubling condition. In Chapter III we discuss the properties of the sharp maximal functions as well as those of the so-called local sharp maximal functions. This is also done in the context of homogeneous spaces, and the results proved include an extension of the John-Nirenberg inequality.

In Chapter IV we consider the functions defined on the upper-half space R_+^{n+1} which are of interest to us, including the nontangential maximal function and the area function. Then, in Chapter V, we restrict our attention to a particular class of functions defined on R_+^{n+1}, namely, the extensions of a tempered distribution on R^n to the upper-half space R_+^{n+1} by means of convolutions with the dilates of Schwartz functions. We study how the extension behaves with respect to different Schwartz functions, and an interesting result is the mean-value type inequality we show these extensions satisfy.

We are now ready to introduce the weighted Hardy spaces in Chapter VI. We also describe here some of their essential properties, such as the independence of the "norm" among others. In Chapter VII we construct a dense class of functions for these spaces of distributions; this is a delicate pursuit. Chapters VIII and IX lie at the heart of the matter: in Chapter VIII we construct the atomic decomposition for these spaces, and in Chapter IX we describe an extension of the Fefferman H^1 duality result by means of the so-called basic inequality.

In Chapters X, XI and XII we then discuss some applications. Chapter X contains the construction of the dual to the Hardy spaces, Chapter XI deals with the continuity of various singular integral and multiplier operators on these spaces, and, finally, in Chapter XII we show how the complex method of interpolation applies in these context. All in all, the essential ingredients of the theory of the weighted Hardy spaces is contained in these notes.

Contents

CHAPTER

I

Weights

We begin our exposition in a general setting. Let X be a metric space endowed with a measure μ. If the measure ν is absolutely continuous with respect to μ and if there exists a nonnegative locally integrable function w such that $d\nu(x) = w(x)d\mu(x)$, we say that ν is a weighted measure with respect to μ and that w is a weight. Throughout these notes we assume that all absolutely continuous measures are weighted measures; this is the case if, for instance, μ is σ-finite.

In this chapter we are mainly concerned with various relations between ν and μ, different conditions on w, and the continuity properties of the Hardy-Littlewood maximal operator. The conditions we have in mind for the weights include Muckenhoupt's A_p condition, the reverse Hölder condition RH_r and the doubling condition D_b. Aside from trivial implications these conditions are independent of each other and consequently their detailed study is justified.

Most of the results presented in this chapter are well-known in the case $X = R^n$ and μ the Lebesgue measure. The proofs we present differ methodologically from the usual ones since we don't have at our disposal tools such as the dyadic decomposition, replaced here by the notion of telescoping sequence of balls.

Assume, then, that X is a metric space with measure μ and that the class of compactly supported continuous functions is dense in the space of integrable functions $L(\mu)$. Further, suppose there is a nonnegative real-valued function d in $X \times X$, it need not be the distance function in X, that satisfies the following properties:

(i) $d(x,x) = 0$ for all x in X.
(ii) $d(x,y) > 0$ for all $x \neq y$ in X.
(iii) There is a constant $c_0 \geq 1$ such that $d(x,y) \leq c_0 d(y,x)$ for all x, y in X.

(iv) There is a constant $c_1 > 0$ such that $d(x,y) \leq c_1(d(x,z) + d(y,z))$ for all x, y, z in X.

(v) For each neighbourhood $\mathcal{N}$ of x in X there is an $r > 0$ such that the ball $B(x,r) = \{y \in X : d(x,y) \leq r\}$ is contained in $\mathcal{N}$.

(vi) The balls $B(x,r)$ are measurable for all $x \in X$ and $r > 0$.

(vii) (Doubling D_b condition) There are a constant $k > 0$ and a number $b > 0$ with the property that for all $x \in X$, $t \geq 1$ and $r > 0$ we have

$$\mu(B(x,tr)) \leq kt^b \mu(B(x,r)). \tag{D_b}$$

We write $\mu \in D_b$ to indicate that μ satisfies the D_b condition.

The family of balls satisfies the following geometric property.

Lemma 1. Let $a > 0$. Then there is a constant $c_2 = c_1^2(1+a) + c_0c_1a$ such that if $B(x,r) \cap B(y,r') \neq \varnothing$ and $r \leq ar'$, then $B(x,r) \subseteq B(y,c_2r')$.

Proof. Let $z \in B(x,r)$ and $z_1 \in B(x,r) \cap B(y,r')$. Then

$$d(y,z) \leq c_1(d(y,x) + d(z,x)) \leq c_1(c_1d(y,z_1) + c_1d(x,z_1)) + c_0c_1d(x,z)$$
$$\leq c_1^2r' + c_1^2r + c_0c_1r \leq c_1^2r' + c_1^2ar' + c_0c_1ar' = c_2r',$$

that is, $z \in B(y,c_2r')$. ∎

Using this geometric property it is not difficult to obtain a covering lemma.

Lemma 2. Let $\mathcal{F}$ be a family $\{B(x,r)\}$ of balls with bounded radii. Then there is a countable subfamily $\{B(x_i,r_i)\}$ consisting of pairwise disjoint balls such that each ball in $\mathcal{F}$ is contained in one of the balls $B(x_i,cr_i)$, where c corresponds to the choice $a = 2$ in Lemma 1.

Proof. Let M be a bound for the radii of the balls. For each $n = 0, 1, 2, \ldots$ we construct inductively a family $\{B(x_{i,n},r_{i,n})\}$ of balls with the following properties:

1. $B(x_{i,n},r_{i,n}) \in \mathcal{F}$, and $2^{-n-1}M < r_{i,n} \leq 2^{-n}M$.

2. The $B(x_{i,m},r_{i,m})$'s are pairwise disjoint for $0 \leq m \leq n$.

3. For each n the family is maximal with respect to 1. and 2.

Let now $B(x,r) \in \mathcal{F}$. If $2^{-n-1}M < r \leq 2^{-n}M$, then $B(x,r)$ intersects one of the balls $B(x_{i,m},r_{i,m})$ with $m \leq n$. In this case it follows that $r < 2r_{i,m}$ and, by Lemma 1, $B(x,r) \subseteq B(x_{i,m},cr_{i,m})$. ∎

Given a locally integrable function f we define the Hardy-Littlewood maximal function $M_\mu f$ of f by

$$M_\mu f(x) = \sup_{B(y,r) \supset \{x\}} \frac{1}{\mu(B(y,r))} \int_{B(y,r)} |f| \, d\mu.$$

It is of interest to decide when the maximal function is of weak-type or of type (p,p) on $L^p(\mu)$.

Theorem 3.   The Hardy-Littlewood maximal function M_μ is of weak-type $(1,1)$ and of type (p,p) for $1 < p \leq \infty$ on $L^p(\mu)$.

Proof.   The statement concerning the continuity in $L^p(\mu)$ follows by interpolation from the weak-type $(1,1)$ and the boundedness of M_μ in $L^\infty(\mu)$. As for the weak-type $(1,1)$, for each $n > 0$ let

$$M_{\mu,n}f(x) = \sup_{B(y,r)\supset\{x\},r\leq n} \frac{1}{\mu(B(y,r))} \int_{B(y,r)} |f|\,d\mu\,.$$

We begin by showing that the mapping $M_{\mu,n}$ is of weak-type $(1,1)$ with norm bounded independently on n; once this is done the result follows by letting n tend to ∞. Given $\lambda > 0$, let $\mathcal{F}$ be the family of balls $B = B(y,r)$, $r \leq n$, such that $\lambda\mu(B) < \int_B |f|\,d\mu$ and let $\{B(x_i,r_i)\}$ be the pairwise disjoint family of balls corresponding to $\mathcal{F}$ obtained in Lemma 2. It then follows that

$$\{M_\mu f > \lambda\} \subseteq \bigcup_{B\in\mathcal{F}} B \subseteq \bigcup_i B(x_i, cr_i)\,,$$

and consequently, $\sum_i \mu(B(x_i, r_i)) \leq \lambda^{-1} \int_X |f|\,d\mu$. The desired estimate is obtained now since, by the doubling condition, $\mu(B(x_i, cr_i)) \leq kc^b\mu(B(x_i, r_i))$. ∎

Let now ν be another measure defined on X. We are interested in studying when the Hardy-Littlewood maximal operator M_μ is of weak-type or of type (p,p) on $L^p(\nu)$. Specifically, we search for conditions on ν for the estimates

$$\nu(\{M_\mu f > \lambda\}) \leq c_p\lambda^{-p}\|f\|_{L^p(\nu)}^p\,, \quad \text{all } \lambda > 0\,, \tag{1}_p$$

and

$$\|M_\mu f\|_{L^p(\nu)} \leq c_p\|f\|_{L^p(\nu)}\,. \tag{2}_p$$

Now, if χ_E denotes the characteristic function of a set $E \subseteq B$, we readily see that $M_\mu\chi_E(x) \geq \mu(E)/\mu(B)$ for all $x \in B$, and if $(1)_p$ holds we get

$$\nu(E)/\nu(B) \geq c_p^{-1}(\mu(E)/\mu(B))^p\,, \quad \text{all } E \subseteq B\,. \tag{3}_p$$

Note that if $\mu \in D_b$ and $(3)_p$ holds, then $\nu \in D_{pb}$. Furthermore, $(3)_p$ implies that there is a constant c such that

$$M_\mu\chi_E(x) \leq cM_\nu\chi_E(x)^{1/p}\,, \quad \text{all } x \in X\,.$$

Whence, by the type (p,p) statement in Theorem 3 for M_ν, we conclude that $(3)_p$ implies $(1)_p$ for characteristic functions of sets. Thus we have shown

4

Lemma 4. Condition $(3)_p$ is necessary and sufficient for the restricted weak-type (p,p) inequality for the operator M_μ in $L^p(\nu)$.

Another interesting consequence of $(3)_p$ is

Lemma 5. Condition $(3)_p$ implies that ν is absolutely continuous with respect to μ.

The proof of this lemma will be presented later on, as part of the implication (6) implies (2) in Theorem 15.

Now, assuming for the moment that Lemma 5 has been proved, if $(1)_p$ holds, then ν is a weighted measure with respect to μ with weight w, say. Suppose $p > 1$, let B be an arbitrary ball and let χ_ε denote the characteristic function of the set $\{x \in B : w(x) > \varepsilon\}$. Then for the function $f = w^{-1/(p-1)}\chi_B\chi_\varepsilon$ we have

$$M_\mu f(x) \geq \frac{1}{\mu(B)} \int_B \chi_\varepsilon w^{-1/(p-1)} d\mu, \quad \text{all } x \in B,$$

and from $(1)_p$ it follows that

$$\nu(B) \left(\frac{1}{\mu(B)} \int_B \chi_\varepsilon w^{-1/(p-1)} d\mu \right)^p \leq c^p \int_B \chi_\varepsilon w^{-p/(p-1)} d\nu$$

$$= c^p \int_B \chi_\varepsilon w^{-1/(p-1)} d\mu.$$

Sorting this inequality out and letting ε go to 0 we obtain the so-called $A_p(\mu)$ condition for w, to wit,

$$\frac{1}{\mu(B)} \int_B w\, d\mu \left(\frac{1}{\mu(B)} \int_B w^{-1/(p-1)} d\mu \right)^{p-1} \leq c, \quad \text{all } B. \qquad (A_p(\mu))$$

When this condition holds we write $w \in A_p(\mu)$.

On the other hand, by Hölder's inequality we also have

$$\frac{1}{\mu(B)} \int_B |f|\, d\mu$$

$$\leq \frac{1}{\mu(B)} \left(\int_B |f|^p w\, d\mu \right)^{1/p} \left(\int_B w^{-1/(p-1)} d\mu \right)^{(p-1)/p}$$

$$= \left(\frac{1}{\nu(B)} \int_B |f|^p d\nu \right)^{1/p} \left(\frac{1}{\mu(B)} \int_B w\, d\mu \left(\frac{1}{\mu(B)} \int_B w^{-1/(p-1)} d\mu \right)^{p-1} \right)^{1/p}.$$

Whence, if $w \in A_p(\mu)$, it follows that $M_\mu f(x) \leq cM_\nu f(x)^{1/p}$, and from Theorem 3 we conclude that $(1)_p$ holds. We have thus shown, assuming Lemma 5 of course,

Lemma 6. Let $1 < p < \infty$. Then, M_μ is of weak-type (p,p) on $L^p(\nu)$ if and only if ν is a weighted measure with respect to μ, and in this case the weight $w \in A_p(\mu)$.

As for the case $p = 1$, assume $(1)_1$ holds and put $f = \chi_B(1 - \chi_\varepsilon)/\varepsilon$. Then, if $E_\varepsilon = \{x \in B : w(x) \le \varepsilon\}$, we have $M_\mu f(x) \ge \mu(E_\varepsilon)/(\varepsilon\mu(B))$ for $x \in B$, and $\|f\|_{L^1(\nu)} \le \mu(E_\varepsilon)$. These observations give

$$\mu(E_\varepsilon)(\nu(B)/\varepsilon\mu(B)) \le c\mu(E_\varepsilon).$$

Whence, $\mu(E_\varepsilon) = 0$ when $\nu(B)/(\varepsilon\mu(B)) > c$, and the condition $A_1(\mu)$ holds, i.e.,

$$\left(\operatorname*{ess\ inf}_{x \in B} w(x)\right)^{-1} \frac{1}{\mu(B)} \int_B w \, d\mu \le c, \quad \text{all } B. \qquad (A_1(\mu))$$

When this condition is satisfied we write $w \in A_1(\mu)$.

Note that if w is a nonnegative measurable function and $d\nu = w d\mu$ we have

$$\frac{1}{\mu(B)} \int_B |f| \, d\mu \le \left(\frac{1}{\nu(B)} \int_B |f| \, d\nu\right) \left(\operatorname*{ess\ inf}_{x \in B} w(x)\right)^{-1} \left(\frac{1}{\mu(B)} \int_B w \, d\mu\right),$$

and consequently, if $w \in A_1(\mu)$ we get $M_\mu f(x) \le c M_\nu f(x)$ for all x. In this case, by Theorem 3 we conclude that M_μ is of weak-type $(1,1)$ on $L^1(\nu)$. Summing up, we have proved

Lemma 7. M_μ is of weak-type $(1,1)$ on $L^1(\nu)$ if and only if ν is a weighted measure with respect to μ and the weight $w \in A_1(\mu)$.

Returning to the $A_p(\mu)$ condition, one of its basic features is

Lemma 8. If $p > 1$ and $w \in A_p(\mu)$, there is an $\varepsilon > 0$ so that $w \in A_{p-\varepsilon}(\mu)$.

Note that if $w \in A_p(\mu)$, then by Hölder's inequality it follows that $w \in A_q(\mu)$ for $q > p$. On the other hand, Lemma 8, proved after Theorem 18, is far from trivial. It is a crucial property in the theory of weights and it implies the maximal theorem.

Theorem 9. Let $p > 1$. Then the inequality $\|M_\mu f\|_{L^p(\nu)} \le c_p \|f\|_{L^p(\nu)}$ holds for every $f \in L^p(\nu)$ if and only if ν is a weighted measure with respect to μ and the weight $w \in A_p(\mu)$. In this case, M_μ is of weak-type (p,p) on $L^p(\nu)$ if and only if it is of type (p,p) on $L^p(\nu)$.

Proof. It only remains to prove the sufficiency. Since $M_\mu f(x) \leq c M_\nu f(x)^{1/(p-\varepsilon)}$ when $w \in A_{p-\varepsilon}(\mu)$ and since $p/(p-\varepsilon) > 1$, it follows from Theorem 3. ∎

A close relation to $A_p(\mu)$ is the reverse Hölder condition $RH_r(\mu)$, $r \geq 1$, requiring that for some constant $c > 0$ we have

$$\left(\frac{1}{\mu(B)} \int_B w^r \, d\mu \right)^{1/r} \leq c \left(\frac{1}{\mu(B)} \int_B w \, d\mu \right), \quad \text{all } B. \qquad (RH_r(\mu))$$

When this happens we write $w \in RH_r(\mu)$.

Note that if μ and ν are mutually absolutely continuous and $d\nu(x) = w(x)d\mu(x)$, then the assumption $w \in RH_r(\mu)$ is equivalent to $(1/w) \in A_p(\mu)$, where $(r-1)(p-1) = 1$.

Now for the case $p = \infty$ and the condition $A_\infty(\mu)$. For each ball B we define the median value $w_{B,\mu}$ of the weight w with respect to the measure μ by the expression

$$w_{B,\mu} = (t_1 t_2)^{1/2},$$

where

$$t_1 = \sup\{t > 0 : \mu(\{x \in B : w(x) < t\}) \leq \mu(B)/2\}$$

and

$$t_2 = \inf\{t > 0 : \mu(\{x \in B : w(x) > t\}) \leq \mu(B)/2\}.$$

Observe that for any real number a we have

$$(w^a)_{B,\mu} = (w_{B,\mu})^a, \quad \text{and} \quad \int_B w^a \, d\mu / \mu(B) \geq (w_{B,\mu})^a / 2.$$

We say that $w \in A_\infty(\mu)$ if

$$\frac{1}{\mu(B)} \int_B w \, d\mu \leq c w_{B,\mu}, \quad \text{all } B. \qquad (A_\infty(\mu))$$

A computation using the above observations gives that if $0 < a \leq 1$ and $w \in A_\infty(\mu)$, then

$$\frac{1}{\mu(B)} \int_B w^a \, d\mu \sim (w_{B,\mu})^a, \quad \text{all } B.$$

The next two lemmas describe $A_p(\mu)$ weights in terms of the $A_\infty(\mu)$ condition.

Lemma 10. Let $1 < p < \infty$. Then $w \in A_p(\mu)$ if and only if $w \in A_\infty(\mu)$ and $w^{-1/(p-1)} \in A_\infty(\mu)$.

Proof. The necessity follows from the inequalities

$$\frac{1}{\mu(B)} \int_B w \, d\mu \geq \frac{1}{2} w_{B,\mu} = \frac{1}{2} ((w^{-1/(p-1)})_{B,\mu})^{-(p-1)}$$

and

$$\frac{1}{\mu(B)} \int_B w^{-1/(p-1)} d\mu \geq \frac{1}{2} (w_{B,\mu})^{-1/(p-1)}.$$

As for the sufficiency, it follows since both inequalities above may be reversed with an appropriate constant on the right-hand side. ∎

Lemma 11. Let $1 < p, r < \infty$. Then $w \in A_p(\mu) \cap RH_r(\mu)$ if and only if $w^r \in A_\infty(\mu)$ and $w^{-1/(p-1)} \in A_\infty(\mu)$.

Proof. The necessity follows from the inequality

$$\frac{1}{\mu(B)} \int_B w^r d\mu \leq c \left(\frac{1}{\mu(B)} \int_B w \, d\mu \right)^r \leq c(w_{B,\mu})^r = c(w^r)_{B,\mu}.$$

The sufficiency follows from the inequalities

$$\frac{1}{\mu(B)} \int_B w \, d\mu \leq \left(\frac{1}{\mu(B)} \int_B w^r d\mu \right)^{1/r} \leq c(w^r)_{B,\mu}^{1/r} = c w_{B,\mu}$$

and

$$\frac{1}{\mu(B)} \int_B w^r d\mu \leq c(w_{B,\mu})^r \leq c \left(2 \frac{1}{\mu(B)} \int_B w \right)^r.$$

Also, note that the lemma is equivalent to $w^a \in A_\infty(\mu)$ for all a with $-1/(p-1) \leq a \leq r$. ∎

For the remainder of this chapter we no longer automatically assume that μ satisfies a doubling condition. Our next aim is to show that the eight conditions we list below are, under some additional assumptions on the measure μ and the balls $B(x,r)$, equivalent ways of stating the fact that $w \in A_p(\mu)$ for some p, $1 \leq p < \infty$. The conditions are:

(1) There are $0 < \varepsilon, \delta < 1$ such that for each ball B and each measurable set $E \subseteq B$, we have $\nu(E) < (1 - \delta)\nu(B)$ whenever $\mu(E) < \varepsilon \mu(B)$.

(2) There are a constant $c > 0$ and an index $p \geq 1$ such that $\nu(E)/\nu(B) \leq c(\mu(E)/\mu(B))^{1/p}$ for each ball B and each measurable set $E \subseteq B$.

(3) ν is a weighted measure with respect to μ and the weight w satisfies the condition $RH_r(\mu)$ for some $r > 1$.

(4) ν is a weighted measure with respect to μ and there is a constant c such that $w_{B,\mu} \leq c\nu(B)/\mu(B)$ for all balls B.

8

(5) ν is a weighted measure with respect to μ and there is a constant c such that $w_{B,\nu} \leq cw_{B,\mu}$ for all balls B.

(6) ν is a weighted measure with respect to μ and there are a constant c and an index $p \geq 1$ so that $\nu(E)/\nu(B) \geq c(\mu(E)/\mu(B))^p$ for each ball B and each measurable set $E \subseteq B$.

(7) ν is a weighted measure with respect to μ and the weight w satisfies the condition $A_p(\mu)$ for some $p \geq 1$.

(8) ν is a weighted measure with respect to μ and $\nu(B)/\mu(B) \leq cw_{B,\mu}$ for all balls B.

First some observations. Note that condition (1) is symmetric in ν and μ, and that (2) becomes (6) if we exchange ν and μ. Also, if ν and μ are mutually absolutely continuous, then (5) is symmetric in μ and ν, and if we exchange μ and ν (3) becomes (7) and (4) becomes (8).

Now, without assuming a doubling condition on either μ or ν, we have

Lemma 12. The following diagram is true:

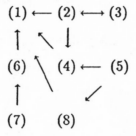

Also, (2) together with (6) imply both (7) and (8), while (4) together with (8) imply (5).

Note that (2) together with (6) imply every other condition and that (1) is implied by every other condition. However, there is no implication emanating from (1). The crucial step in proving the equivalence of the above conditions is (1) implies (2). In this direction we have

Theorem 13. Suppose that μ and ν satisfy a doubling condition. Then (1) implies (2) and, by symmetry, (1) implies (6).

Corollary 14. Under the assumption that μ and ν satisfy a doubling condition, (1)-(8) are equivalent.

On the other hand, if we only assume the doubling condition for μ, (6) implies a doubling condition for ν, and by the (6) $\rightarrow$ (1) and Theorem 13 it also follows that (6) $\rightarrow$ (2). This makes it possible to complete the results of Lemma 12 and obtain

Theorem 15. Suppose μ satisfies a doubling condition. Then the following diagram holds:

$$(3) \longleftrightarrow (2) \longrightarrow (4) \longrightarrow (1)$$
$$\uparrow \qquad \uparrow \qquad \uparrow$$
$$(7) \longleftrightarrow (6) \longrightarrow (5) \longrightarrow (8)$$

By means of an example we show below that without further assumptions the above diagram cannot be improved. Nevertheless, if $X = R^n$, μ is doubling and the balls are the dyadic cubes of R^n, the doubling condition on ν is not required in Theorem 13 for the implication $(1) \rightarrow (2)$. Indeed, we show that the following holds.

Theorem 16. Let $X = R^n$, and suppose μ satisfies a doubling condition and the B's are the dyadic cubes of R^n. Then the following implications are true:

$$(7) \longleftrightarrow (6) \longrightarrow (5) \longleftrightarrow (8) \longrightarrow (4) \longleftrightarrow (3) \longleftrightarrow (2) \longleftrightarrow (1)$$

To see that this result cannot be improved, let $X = R$ with the dyadic intervals $(k2^j, (k+1)2^j)$ and put

$$\rho_0(x) = \begin{cases} 0 & \text{if } 0 < x < 1 \\ 1 & \text{otherwise}, \end{cases} \qquad \rho_1(x) = \begin{cases} 1/2 & \text{if } 0 \le x \le 1 \\ -1/2 & \text{if } 1 < x \le 2 \\ 0 & \text{otherwise}. \end{cases}$$

Now, if $d\mu(x) = (1 + \rho_1(x))dx$ and $d\nu(x) = \rho_0(x)dx$, then (1)-(4) are satisfied, but (5)-(8) are not. If, on the other hand, $d\mu(x) = (1 - \rho_1(x))dx$, then (1)-(5) and (8) are satisfied, but not (6) or (7).

It is possible to avoid assuming that ν is doubling in Theorem 13, provided some restrictions are impossed on the family of balls. Specifically, we have

Theorem 17. Suppose that μ satisfies a doubling condition and that the function $\mu(B(x,r))$ increases continuously with r for each $x \in X$. Then, condition (1) implies that ν also satisfies a doubling condition.

From Theorems 13 and 17 we conclude the following result.

Theorem 18. Suppose that μ satisfies a doubling condition and that $\mu(B(x,r))$ increases continuously with r for each $x \in X$. Then conditions (1)-(8) are equivalent and μ and ν are mutually absolutely continuous.

The last statement of Theorem 17 follows directly from (2) and (6).

We pass now to prove Lemma 12; the proofs of Theorems 13 and 17 are given after Lemma 24.

Proof of Lemma 12. $(2) \to (1)$ is immediate. $(6) \to (1)$ is also immediate once we observe that (1) is symmetric in μ and ν. $(3) \to (2)$ and $(7) \to (6)$ are a straightforward application of Hölder's inequality to the expressions $\int_B \chi_E w \, d\mu$ and $\int_B \chi_E w^{1/p} w^{-1/p} d\mu$ respectively. As for the implication $(2) \to (3)$, the statement concerning the absolute continuity is immediate. Let now $E_\lambda = \{x \in B : w(x) > \lambda\}$. By Chebychev's inequality and (2) it follows that $\lambda \mu(E_\lambda) \leq \nu(E_\lambda) \leq \nu(B)(\mu(E_\lambda)/\mu(B))^{1/p}$. Thus $\mu(E_\lambda) \leq \min(\mu(B), \nu(B)^{p'}/(\lambda \mu(B)^{1/p})^{p'}$, which, substituted into the expression $\int_B w^r d\mu = r \int_0^\infty \lambda^{r-1} \mu(E_\lambda) \, d\lambda$, gives (3) with $r < 1 + 1/p$.

$(8) \to (1)$. Let $\varepsilon = 1/4$. If $\mu(E) \leq \mu(B)/4$, then it follows that

$$\mu(\{x \in B \setminus E : w(x) \geq w_{B,\mu}\}) \geq c\mu(B) \,.$$

Thus $\int_{B \setminus E} w \, d\mu > 4 w_{B,\mu} \mu(B) \geq \nu(B)/c$, and so $\nu(E) < (1 - (1/c))\nu(B)$.

$(4) \to (1)$. Observe that $\nu(E) \leq \nu(B)/2 + \nu(\{x \in E : w(x) < w_{B,\mu}\}) \leq \nu(B)(1/2 + (\mu(E)/\mu(B)))$.

$(2) \to (4)$. The absolute continuity is immediate. Let now $E = \{x \in B : w(x) > w_{B,\mu}\}$; then we have $\nu(B) \geq 2\nu(E) \geq 2w_{B,\mu}\mu(E)$. But by (2), $\mu(E)/\mu(B) > (\nu(E)/\nu(B))^p/c > 0$, and so $w_{B,\mu}\mu(B) < c\nu(B)$.

$(5) \to (4)$. Let $E = \{x \in B : w(x) \geq w_{B,\mu}\}$. Then we get that $\mu(B) \leq 2\mu(E) \leq 2 \int_E w \, d\mu/w_{B,\mu} \leq \mu(B)/w_{B,\nu}$, from which (4) follows.

$(5) \to (8)$. Let $E = \{x \in B : w(x) \leq w_{B,\nu}\}$. Then $\nu(B) \leq 2\nu(E) = 2 \int_E w \, d\mu \leq 2\mu(E)w_{B,\mu} < c\mu(B)w_{B,\mu}$ from which (8) follows.

That (4) and (8) imply (5) is immediate. We now show that (2) and (6) imply (7). From (2) we conclude that ν is absolutely continuous with respect to μ, and from (6) that μ is absolutely continuous with respect to ν. Thus, $d\mu(x) = (1/w(x))d\nu(x)$ and, by (6), $(1/w)$ satisfies a reverse Hölder inequality with respect to ν. From $(3) \to (2)$ it now follows that (7) holds. Finally we show that (2) and (6) imply (8). As above we get that μ and ν are mutually absolutely continuous and that that $d\mu(x) = (1/w(x))d\nu(x)$. Also, (6) implies that (2) holds with μ and ν exchanged. Since $(1/w)_{B,\mu} = 1/w_{B,\mu}$, (8) follows from the implication $(2) \to (4)$. ∎

Lemmas 5 and 8 follow from Theorem 15. Indeed, Lemma 5 follows from $(6) \to (2)$ in that theorem, and we now indicate how $(7) \to (2)$ there gives Lemma 8. First note that if $w \in A_p(\mu)$, then $w^{-1/(p-1)} \in A_{p_1}(\mu)$ for some $p_1 > 1$. Thus, $w^{-1/(p-1)} \in RH_r(\mu)$ for some $r > 1$, i.e.,

$$\frac{1}{\mu(B)} \int_B w^{-r/(p-1)} d\mu \leq c \left(\frac{1}{\mu(B)} \int_B w^{-1/(p-1)} d\mu \right)^r, \quad \text{all } B \,.$$

From this inequality and the fact that $w \in A_p(\mu)$ we conclude that $w \in A_{p-\varepsilon}(\mu)$ where $1/(p - \varepsilon - 1) = r/(p-1)$.

From Lemma 7 it follows that the set of values of $p > 1$ for which $w \in A_p(\mu)$ is an open interval (p_w, ∞) or the empty set, i.e., $p_w = \infty$. The number p_w is called the critical index of the weight w for the $A_p(\mu)$ condition. Also, if $w \in A_p(\mu) \cap RH_r(\mu)$, $p, r > 1$, from the observation that $w^r \in A_{p_1}(\mu)$ for some $p_1 > 1$ it readily follows that $w \in A_{p-\varepsilon}(\mu) \cap RH_{r+\varepsilon}(\mu)$ for some $\varepsilon > 0$. Thus the set of values of $r > 1$ for which $w \in RH_r(\mu)$ is an open interval $(1, r_w)$ for some number $r_w > 1$ provided that w is in some $A_p(\mu)$ class. The number r_w is called the critical index of the weight w for the $RH_r(\mu)$ condition. By the way, the condition $w \in RH_r(\mu)$ alone does not necessarily imply that $w \in RH_{r+\varepsilon}(\mu)$ for some $\varepsilon > 0$ unless we are under the assumptions of Theorem 18.

In view of Lemmas 10 and 11 we formulate the above remarks as follows:

Lemma 19. Let w be a locally integrable weight, and let I denote the set of those real numbers a such that

$$\frac{1}{\mu(B)} \int_B w^a \, d\mu \leq c(w_{B,\mu})^a.$$

Then one of the following three conditions holds:
   (i) I is empty,
   (ii) I is an interval with 0 as an endpoint,
   (iii) I is an open interval (a_-, a_+) with $a_- < 0$ and $a^+ > 0$. In this case $a_- = -(p_w - 1)^{-1}$ and $a_+ = r_w$, where p_w and r_w are the critical indices of w.

Furthermore, note that under the assumptions of Theorem 18, (ii) cannot occur.

Before we proceed to prove Theorems 13 and 17 we introduce a "reverse" doubling condition. We say that μ satisfies the reverse doubling condition RD_d, and we write $\mu \in RD_d$, if there are a nonnegative constant $k_1 \leq 1$ and a number $d > 0$ such that

$$\mu(B(x,tr)) > k_1 t^d \mu(B(x,r)), \quad \text{all } t \geq 1 \text{ and } \quad B(x,r). \qquad (RD_d)$$

In case X is compact we only require (RD_d) for $tr \leq c$.

Lemma 20. Assume that there is a constant $c_3 > 1$ such that for all x in X and $r > 0$, or $0 < r < c$ if X is compact, we have $B(x, c_3 r) \setminus B(x,r) \neq \emptyset$. If $\mu \in D_d$ for some $d \geq 1$, then also $\mu \in RD_{d_1}$ for some $d_1 < 1$.

Proof. The conclussion is apparent if there are constants $a_1 > 1$ and $0 < a_2 < 1$ such that

$$\mu(B(x,r)) \leq a_2 \mu(B(x, a_1 r)), \quad \text{all } B(x,r).$$

In case X is compact we only need this inequality for $0 < r < c$. As for the inequality itself, it follows from D_d provided we can find a constant $c_4 > 1$ with the following property: For each $B(x,r)$ there exists a ball $B(y,r_1) \subseteq B(x,a_1 r)$ such that $B(x,r) \cap B(y,r) = \emptyset$ and $B(x,r) \subseteq B(y,c_4 r_1)$. To see that this is the case, put $a = 1$ in Lemma 1, pick $c_4 > c_0 c_2$, and let $y \in B(x,c_3 c_4 r) \backslash B(x,c_4 r) \neq \emptyset$. It then follows that $c_4 r < d(x,y) \leq c_3 c_4 r$, and, by (iii), that $(c_4 r/c_0) < d(y,x) \leq c_0 c_3 c_4 r$. Whence $d(y,x) > c_2 r$, and by Lemma 1 we have, on the one hand, that $B(x,r) \cap B(y,r) = \emptyset$, and, on the other hand, that $B(x,r) \subseteq B(y,a_1 r)$, where $a_1 = c_0 c_2 c_3 c_4$. ∎

It is important to note that there is no implication from RD_{d_1} to D_d. Also, contrary to the $A_p(\mu)$ condition, the set of values of d for which $\mu \in D_d$ can either be a closed interval $[d_w, \infty)$ or an open interval (d_w, ∞) or the empty set, i.e., $d_w = \infty$. d_w is called the critical value of μ for the doubling condition D_d.

A close relation to doubling is the condition B_λ for a measure μ. We say that $\mu \in B_\lambda$ provided that

$$\sum_{k=0}^{\infty} 2^{-\lambda k} \mu(B(x,2^k r)) \leq c\mu(B(x,r)), \quad \text{all } B(x,r). \tag{B_λ}$$

Note that if $\mu \in D_{\lambda-\varepsilon}$ for some $0 < \varepsilon < \lambda$, then $\mu \in B_\lambda$. Also, if $\mu \in B_\lambda$, then $\mu \in D_{\lambda-\varepsilon}$ for some $\varepsilon > 0$. Furthermore, we have

Lemma 21. If $\mu \in B_\lambda$ for some $\lambda > 0$, then there is $0 < \varepsilon < \lambda$ such that $\mu \in B_{\lambda-\varepsilon}$.

Proof. For $k = 0,1,\ldots$ put $a_k = 2^{-\lambda k} \mu(B(x,2^k r))$. Replacing r by $2^{k_0} r$ in the condition B_λ we get

$$\sum_{k=0}^{\infty} a_{k+k_1} \leq c a_{k_1}.$$

By taking next a sum over $k_1 \geq k_2$ we get

$$\sum_{k_1=0}^{\infty} \sum_{k=0}^{\infty} a_{k+k_1+k_2} \leq c^2 a_{k_2}.$$

Repeating this argument m times it follows that

$$\sum_{k,k_1,\ldots,k_m} a_{k+k_1+\cdots+k_m} \leq c^{m+1} a_0.$$

Now, given an integer ℓ, let $N_m(\ell)$ denote the number of $(m+1)$-tuples of non-negative integers $k, k_1, \ldots, k_m$ such that $k + k_1 + \cdots + k_m = \ell$. The above estimate may then be rewritten

$$\sum_{\ell} N_m(\ell) a_\ell \leq c^m a_0.$$

Let $0 < \varepsilon < 1/c$. Multiplying the above inequality through by ε^m and summing over $m \geq 0$ it follows that

$$\sum_m \sum_\ell \varepsilon^m N_m(\ell) a_\ell \leq c a_0/(1 - \varepsilon c).$$

But clearly $N_m(\ell) \geq \ell^m/m!$ and so we obtain

$$\sum_{m=0}^{\infty} \varepsilon^m N_m(\ell) \geq e^{\varepsilon \ell}.$$

Whence we conclude that

$$\sum_\ell e^{\varepsilon \ell} a_\ell \leq c' a_0.$$

From this we get that, as asserted, $\mu \in B_{\lambda - \varepsilon}$ for some $\varepsilon > 0$. ∎

Given two measures μ and ν on X, we consider next the relative doubling condition

$$\nu(B_2)/\nu(B_1) \leq c(\mu(B_2)/\mu(B_1))^d, \quad \text{all balls } B_1 \subseteq B_2,$$

where $c \geq 1$ and $d \geq 1$ are constants independent of the balls, and the relative reverse doubling condition

$$\nu(B_2)/\nu(B_1) \geq c(\mu(B_2)/\mu(B_1))^{d_1}, \quad \text{all balls } B_1 \subseteq B_2,$$

where $c > 0$ and $0 < d_1 < 1$ are constants, also independent of the balls.

If $\mu(B(x,r)) \sim r$, the relative doubling condition essentially reduces to $\nu \in D_d$ and the relative reverse doubling inequality to $\nu \in RD_{d_1}$.

We now show that under some additional assumptions on X and μ, it is possible to find a normalized distance function $d^*(x,y)$ with the property that the collection $\{B^*(x,r)\}_{r>0}$ of balls relative to d^* coincides with $\{B(x,r)\}_{r>0}$ for each $x \in X$ but with the corresponding radii relabeled and so that $\mu(B^*(x,r)) \sim r$. For this purpose we construct for each $x \in X$ a function $r_x: [0,\infty) \to [0,\infty)$ that satisfies the following five properties:

(i') $r_x(0) = 0$.
(ii') r_x is continuous.
(iii') r_x is strictly increasing.
(iv') If X is not compact, then $r_x(t) \to \infty$ as $t \to \infty$.
(v') There is a constant c such that $r_x(t)/c \leq \mu(B(x,t)) \leq c r_x(t)$.

If such a function is available, then $d^*(x,y)$ is plainly defined to be $r_x(d(x,y))$. In this direction we have

Lemma 22. Suppose μ is a doubling measure defined on X. With r_x and $d^*(x,y)$ defined as above, $d^*(x,y)$ satisfies the conditions (i)-(vii) of the function $d(x,y)$ introduced at the beginning of the chapter, the collection of balls $\{B^*(x,r)\}_{r>0}$ coincides with $\{B(x,r)\}_{r>0}$, and $\mu(B^*(x,r)) \sim r$.

Proof. Clearly (i') implies (i) and (iii') implies (ii). As for (iii), first note that

$$d^*(x,y) = r_x(d(x,y)) \le r_x(c_0 d(y,x))$$
$$\le c\mu(B(x,c_0 d(y,x))) \le ckc_0^b \mu(B(x,d(y,x))).$$

Further, if $r = 2d(y,x)$, then $B(x,r) \cap B(y,r) \ne \emptyset$, and by Lemma 1 $B(x,r) \subseteq B(y,c_2 r)$. Thus we may continue with the above estimate and obtain

$$d^*(x,y) \le ckc_0^b \mu(B(y,c_2 d(y,x))) \le ck^2 (c_0 c_2)^b \mu(B(y,d(y,x)))$$
$$\le (ck)^2 (c_0 c_2)^b r_y(d(y,x)) = c_0' d^*(y,x).$$

Next observe that $B(x,t) = B^*(x,s)$ whenever $s = r_x(t)$, and that the inverse function r_x^{-1} is defined on $[0,\infty)$, or on $[0,c_x)$ in case X is compact. Now, in the compact case, if $s \ge c_x$, there is a t such that $B(x,t) = X = B^*(x,r_x(t))$ for $r_x(t) < c_x \le s$, and consequently, $B^*(x,s) = X = B(x,t)$. With these remarks out of the way, properties (iv) and (v) are immediate.

To verify (vi) observe that either $d(x,y) \le 2c_2 d(x,z)$ or $d(x,y) \le c_2 d(y,z)$. In the former case, by the doubling condition, $\mu(B(x,d(x,y))) \le c\mu(B(x,d(x,z)))$ and since $\mu(B(x,t)) \sim r_x(t)$ it follows that $d^*(x,y) \le cd^*(x,z)$. In the latter case, by (iii) we have $d(y,z) \le c_0 d(x,y) \le c_0 c_2 d(y,z)$ and we conclude that $d^*(y,x) \le cd^*(y,z)$; thus $d^*(x,y) \le cd^*(y,z)$. As for (vii), it follows at once from the relation $\mu(B(x,t)) \sim r_x(t)$. ∎

Our next step is to construct the function r_x.

Lemma 23. Assume that μ is a nonidentically zero measure which satisfies both a D_d and an RD_{d_1} condition. Then

$$r_x(s) = e^{-(1+s)^{-1}} \int_{1/2}^{1} \mu(B(x,st))\,dt, \quad s \ne 0, \quad r_x(0) = 0,$$

satisfies the conditions (i')-(v') given above.

Proof. The continuity of r_x away from the origin is obvious and at $s = 0$ follows since, by the RD_{d_1} condition, $\lim_{t \to 0} \mu(B(x,t)) = 0$. This condition also implies in the noncompact case that (iv') holds. r_x is strictly increasing because $\mu(B(x,t))$ is nondecreasing as a function of t and the factor $e^{-(1+s)^{-1}}$ is increasing. As for (v'), it holds since, by the doubling condition, we have $\mu(B(x,t)) \sim \mu(B(x,st))$ whenever $1/2 \le s \le 1$. ∎

The only possible implication between the $A_p(\mu)$ and $B_q(\mu)$ conditions is given by

Lemma 24. Suppose ν is a weighted measure with respect to μ with $w \in A_p(\mu)$, $1 < p$, and that function d is normalized so that $\mu(B(x,r)) \sim r$. Then $\nu \in B_p$.

Proof. Since $w \in A_p(\mu)$ implies that $w \in A_{p-\varepsilon}(\mu)$ for some $\varepsilon > 0$, and since $\nu \in D_{p-\varepsilon}$ implies that $\nu \in B_p$, it is enough to show that $w \in A_p(\mu)$ implies $\nu \in D_p$. Let $B_1 \subseteq B_2$. By Hölder's inequality we get

$$\mu(B_1) = \int_{B_2} \chi_{B_1} w^{-1} d\nu \leq \left(\int_{B_2} \chi_{B_1}^p \, d\nu \right)^{1/p} \left(\int_{B_2} w^{-p/(p-1)} d\nu \right)^{(p-1)/p}$$

$$= \nu(B_1)^{1/p} \left(\int_{B_2} w^{-1/(p-1)} d\mu \right)^{(p-1)/p} \leq c(\nu(B_1)/\nu(B_2))^{1/p} \mu(B_2).$$

Thus $\nu \in D_p$. ∎

We remark that given $1 \leq a \leq b \leq \infty$, it is possible to construct in $X = R$ endowed with the Lebesgue measure μ, a weighted measure ν with respect to μ such that $d\nu(x) = w(x)d\mu(x)$ and the critical indices $p_w = b$ and $d_w = a$.

Finally we present the postponed proofs of Theorems 13 and 17; first some notations and definitions. If $B = B(x,r)$, let $B^1 = B(x,cr)$ where the constant $c \geq c_2$. We say that a sequence $\{B_k\}$ is a telescoping chain of balls if we have

$$B_1^1 \subseteq B_2^1 \subseteq \ldots \subseteq B_k^1 \subseteq \ldots$$

We also say that $\{\mathcal{F}_k\}$ is a telescoping sequence of collections of balls, $\mathcal{F}_k = \{B_{i,k}\}_i$, $k = 1, 2, \ldots$, provided that
1. $B_{i,k} \cap B_{j,k} = \emptyset$, $i \neq j$, for each k.
2. For each $B_{i,k} = B_k \in \mathcal{F}_k$ there is a telescoping chain of balls $B_k^1 \subseteq B_{k+1}^1 \subseteq \ldots$ such that $B_j \in \mathcal{F}_j$ for all $j \geq k$.

Proof of Theorem 17. Suppose that μ is a doubling measure with constant c in the doubling condition D_d, that $\mu(B(x,r))$ is a continuous function of r, and that (1) holds with constants $0 < \varepsilon, \delta < 1$. We claim that for $t > 1$ we have

$$\mu(B(x,r))/\mu(B(x,rt)) \geq 1/ct^d \geq (1-\varepsilon)^k, \quad k \text{ large enough}.$$

In fact, $k = c' + c'' \ln t$, with $c' \sim \ln c / \ln(1/(1-\varepsilon))$ and $c'' = d / \ln(1/(1-\varepsilon))$ will do. Then from the continuity of $\mu(B(x,r))$ it follows that there is a telescoping chain $\{B_j\}$ consisting of $k+1$ balls so that $B_0 = B(x,r)$, $B_k = B(x,rt)$ and $\mu(B_{j-1})/\mu(B_j) \geq (1-\varepsilon)$ for $j = 1, \ldots, k$. Note that for each j this last property of the balls implies that $\mu(B_j \setminus B_{j-1}) \leq \varepsilon \mu(B_j)$, which in turn, by 1., implies that $\nu(B_j \setminus B_{j-1}) \leq (1-\delta)\nu(B_j)$, or $\nu(B_{j-1})/\nu(B_j) \geq \delta$. Thus $\nu(B_0)\delta^{-k} \geq \nu(B_k)$, or, as we wanted to show, $\nu(B(x,rt)) \leq Mt^{d_1}\nu(B(x,r))$. ∎

Proof of Theorem 13.   To prove the assertion we construct a telescoping sequence of collection of balls $\mathcal{F}_k$, $k = 1, 2, \ldots, k_0$, with the following properties: If

$$E_k = \bigcup_{B \in \mathcal{F}_k} B \quad \tilde{E}_0 = E, \quad \text{and} \quad \tilde{E}_k = \bigcup_{B \in \mathcal{F}_k} B^1 \quad \text{for } k \geq 1,$$

then we have $E \subseteq \tilde{E}_1$, $\tilde{E}_{k_0} = B^1$, and for $k \geq 1$ and $B_k \in \mathcal{F}_k$

$$\nu(\tilde{E}_{k-1} \cap B_k) \leq (1 - \beta)\nu(B_k) \quad \text{for some} \quad 0 < \beta < 1. \tag{a}$$

Once such a telescoping sequence of collection of balls has been constructed the conclusion obtains since, with c the doubling constant for ν, it follows that

$$\nu(\tilde{E}_{k-1}) \leq \nu(\tilde{E}_{k-1} \cap E_k) + \nu(\tilde{E}_k \setminus E_k)$$
$$\leq (1 - \beta)\nu(E_k) + \nu(\tilde{E}_k) - \nu(E_k) \leq (1 - (\beta/c))\nu(\tilde{E}_k).$$

In fact, this is the only place in the proof where the doubling condition is invoked. When $X = R^n$ and the balls are the dyadic cubes, then $B^1 = B$ and the doubling condition on ν is not required.

Returning to the proof, from the above inequality we get

$$\nu(E)/\nu(B) \leq c\nu(E)/\nu(\tilde{B}) \leq (1 - (\beta/c))^{k_0},$$

and the desired result follows from this estimate because, as we show below, k_0 is of order $(\ln(\mu(E)/\mu(B))/\ln(c/\varepsilon))$. So, it only remains to construct the $\mathcal{F}_k$'s and to estimate k_0. First observe that there is $r > 0$ such that for $x \in B$, $B(x,r) \subseteq B^1$ and $\mu(B) \leq c_3\mu(B(x,r))$. Thus, with c_4 a constant to be determined, and if $E \subseteq B$ has measure $\mu(E) \leq \varepsilon\mu(B)/(c_3 c_4)$ and $x \in E$, then we can find a ball $B(x,r)$ such that

$$\mu(E \cap B(x,r)) \leq \mu(E) \leq \varepsilon\mu(B)/(c_3 c_4) \leq \varepsilon\mu(B(x,r))/c_4.$$

By the density property of the measure μ, which holds since by assumption functions in $L^1_{\text{loc}}(\mu)$ differentiate the integral μ-a.e., for each $x \in E$ except possibly for a subset of μ measure zero, we can assign a ball $B(x,r_x)$ with $r_x \leq r$ with the property that for some constant $c_5 > 0$

$$\varepsilon\mu(B(x,r_x))/(c_4 c_5) \leq \mu(E \cap B(x,r_x)) \leq \varepsilon\mu(B(x,r_x))/c_4. \tag{b}$$

Let now $\mathcal{F}_1 = \{B_j\}$ be a pairwise disjoint subfamily of these balls such that each $B(x,r_k)$ is contained in B_j^1 for some $B_j \in \mathcal{F}_1$. We now proceed recursively: Having selected $\mathcal{F}_j$ for $j = 1, \ldots, k-1$, if $\mu(E_{k-1}) \leq \varepsilon\mu(B)/(c_3 c_4)$, we construct $\mathcal{F}_k$ in the same way we constructed $\mathcal{F}_1$ but replacing E above by $\tilde{E}_{k-1}$. This process goes on as long as the above inequality holds. That is, if $\mathcal{F}_{k_0}$ is the last collection we construct in this fashion, then we have

$$\mu(\tilde{E}_{k_0}) > \varepsilon\mu(B)/(c_3c_4).$$

We verify that $\{\mathcal{F}_k\}$ is a telescoping sequence of collections of balls. First, by construction the balls in $\mathcal{F}_k$ are pairwise disjoint, $k = 1, 2, \ldots, k_0$. Also, if $B(x,r) \in \mathcal{F}_{k-1}$, there is a ball $B(x,r_x)$ with $r_x \geq r$ so that (a) holds with E replaced by E_{k-1}. Since $B(x,r_x)$ was competing in the covering argument used to select $\mathcal{F}_k$, we have $B(x,r_x) \subseteq B_k^1$ for some $B_k \in \mathcal{F}_k$; to conclude that $B^1(x,r) \subseteq B_k^1$ we need to know whether $r_x > c_0r$. That this is the case follows from the doubling condition since

$$\mu(B(x,r_x)) \geq c_4\mu(E_{k-1} \cap B(x,r_x))/\varepsilon \geq c_4\mu(B(x,r))/\varepsilon$$

provided that we pick c_4 large enough. Since μ and ν satisfy condition (1), from (b) we conclude that (a) holds.

Finally we estimate k_0. Since the balls in $\mathcal{F}_k$ satisfy (b) with E replaced by E_{k-1} there, we get that $\mu(\tilde{E}_k) \leq \mu(E_{k-1})c_4c_5/\varepsilon$. Thus, by the doubling condition,

$$\mu(\tilde{E}_k) \leq \mu(\tilde{E}_{k-1})Mc_4/\varepsilon \quad \text{and} \quad \mu(E_{k_0}) \leq \mu(E)(Mc_4/\varepsilon)^{k_0}.$$

Whence, since $\mu(E_{k_0}) \geq \mu(B)\varepsilon/c_3c_4$, we see that

$$k_0 \geq (\ln(\mu(B)/\mu(E)))/\ln(Mc_4/\varepsilon). \quad \blacksquare$$

Sources and Remarks. That the condition A_p is relevant in the study of the weighted inequalities for the Hardy-Littlewood maximal function is due to B. Muckenhoupt [1972]. The basic properties of the A_p weights were established by B. Muckenhoupt [1974], F. Gehring [1973] and R. Coifman and C. Fefferman [1974]. In the generality given here the theory was considered by A. P. Calderón [1976], whose approach we have adopted, and by R. Coifman and G. Weiss [1977]. Lemma 4 is due to R. Kerman, cf. R. Kerman and A. Torchinsky [1982], and Lemma 10 was established independently by R. Wheeden. Examples showing that aside from the implication in Lemma 24 the various conditions discussed are independent from each other were constructed by B. Muckenhoupt and C. Fefferman [1974] and J.-O. Strömberg [1979b].

CHAPTER

II

Decomposition of Weights

This chapter is devoted to proving a factorization of $A_p(\mu)$ weights with control on the doubling condition they satisfy. We assume that X, μ and d are as in Chapter I and that $\mu(B(x,r)) \sim r$. In addition we assume that there are nontrivial α-Lipschitz functions defined on X; we make this condition precise. Whenever B denotes the ball $B(x,r)$, we let B^k denote the ball $B(x,c^k r)$ where the constant c is so large that every ball with radius $r_1 \leq 2r$ that intersects B is contained in B^1. Further, we assume that c is so large that there are a function $a(x)$ with support contained in B^1 such that $0 \leq a(x) \leq 1$, $a(x) = 1$ for $x \in B$, and a constant M so that $|a(x) - a(y)| \leq M(d(x,y)/r)^\alpha$, $\alpha > 0$. a is called an α-Lipschitz function.

As for the decomposition, it is given in

Theorem 1. Assume ν is a weighted measure with respect to μ with weight $w \in A_p(\mu) \cap RH_r(\mu)$ for some $p, r > 1$. Then we can write $w = w_1^r w_2^{1-p}$ where w_1 and w_2 are $A_1(\mu)$ weights. Furthermore, if $\nu \in RD_{d_1} \cap D_{d_2}$, $0 < d_1 < 1 < d_2$, and if $\varepsilon > 0$, and $d\nu_1(x) = w_1(x)d\mu(x)$ and $d\nu_2(x) = w_2(x)d\mu(x)$, then $\nu_1 \in RD_{d_1-\varepsilon}$ and $\nu_2 \in D_{d_2+\varepsilon}$.

We remark that if $w_1, w_2 \in A_1(\mu)$, then by Hölder's inequality it follows that $w_1 w_2^{1-p} \in A_p(\mu)$. Theorem 1 is a strong converse to this observation.

The proof of Theorem 1 is quite intricate and is achieved in a number of steps. First we decompose w in a large fixed ball B_0, say, and then the general result follows by an easy limiting argument. Also, it is convinient to work with $\phi = \ln w$ instead, to write $\phi = \phi^+ + \phi^-$, say, and then to set $w_1 = e^{\phi^+}$ and $w_2 = e^{\phi^-}$. We build up the functions ϕ^+ and ϕ^- from nonnegative, respectively nonpositive, Lipschitz functions supported on certain collections of "red" and "blue" balls. This is how it goes. Given an arbitrary real-valued function ϕ, let $(e^\phi)_B$ denote the median value of the nonnegative function e^ϕ, and let

$$\phi_B = \ln((e^\phi)_B).$$

ϕ_B is called the median value of the real-valued function ϕ and the meaning of the definition is the following: Whereas the median value of a nonnegative function is given by the geometric mean of two numbers, that of a real-valued function is given by an arithmetic mean instead.

By Lemmas 11 and 19 in Chapter I, the assumption that $w \in A_p(\mu) \cap RH_r(\mu)$ may be restated as follows: Given an arbitrary ball B we have

$$\int_B e^{\beta_1 \phi(x)} d\mu \le c e^{\beta_1 \phi_B} \mu(B), \quad \int_B e^{-\beta_2 \phi(x)} d\mu \le c e^{-\beta_2 \phi_B} \mu(B), \qquad (a)$$

where

$$\beta_1 = r + \varepsilon \quad \text{and} \quad \beta_2 = (1-p)^{-1} + \varepsilon,$$

provided $\varepsilon > 0$ is small enough.

Also, using the facts that $\nu(B)/\mu(B) \sim e^{\phi_B}$ and $\mu(B(x,r)) \sim r$, the assumption $\nu \in RD_{d_1} \cap D_{d_2}$ can be restated as follows: Given arbitrary balls $B_1 = B(x,r_1) \subseteq B_2 = B(y,r_2)$, we have

$$-\frac{1}{\gamma_2}(\ln r_2 - \ln r_1) - c \le \phi_{B_1} - \phi_{B_2} \le \frac{1}{\gamma_1}(\ln r_2 - \ln r_1) + c, \qquad (b)$$

where

$$\gamma_1 = (1 - d_1)^{-1} \quad \text{and} \quad \gamma_2 = (d_2 - 1)^{-1}.$$

Note that by Lemma 22 in Chapter I we may assume that $\gamma_1 \ge \beta_1$ and that $\gamma_2 \ge \beta_2$.

For a small $\delta > 0$, we first write ϕ as the sum of two functions $\phi = \phi^+ + \phi^-$, say, so that (a) and (b) hold with ϕ replaced there by ϕ^+ and ϕ^-, provided that β_i and γ_i are replaced by $\beta_i - \delta$ and $\gamma_i - \delta$, $i = 1, 2$, respectively. In addition, we want ϕ^+ and ϕ^- to satisfy the following condition: Given an arbitrary ball B, we have

$$\phi^+(x) \ge \phi_B^+ - c, \quad \phi^-(x) \le \phi_B^- + c, \quad \text{all } x \in B. \qquad (c)$$

Setting now

$$w_1(x) = e^{(\phi^+(x)/r)} \quad \text{and} \quad w_2(x) = e^{(\phi^-(x)/r)},$$

we have $w = w_1^r w_2^{1-p}$, where w_1 and w_2 satisfy the conditions in the conclusion of Theorem 1.

Note that the first inequality in (c) is equivalent to the fact that ϕ^+ verifies the left-hand side inequality of (b) with $1/\gamma_2 = 0$ and the second inequality is equivalent to the fact that ϕ^- satisfies the right-hand side of (b) with $1/\gamma_1 = 0$.

Let now B_0 be a large fixed ball. We want to show that (a)-(c) hold for all balls contained in B_0.

Let λ be a large positive number. For each integer m we call a ball $B \subseteq B_0^1$ a red ball, or m-red ball, if

$$\phi_B - \phi_{B_0} > (2m+1)\lambda - c_3 ,$$

where c_3 is a large enough constant which is yet to be chosen, independent of B. Similarly, we call a ball $B \subseteq B_0^1$ a blue ball, or m-blue ball, if

$$\phi_B - \phi_{B_0} < (2m-1)\lambda + c_3 .$$

Red balls are often denoted by B_r and blue balls by B_b.

An m-red ball B_r is said to be nested with a ball B if B is larger than B_r, $B_r^2 \cap B^2 \neq \emptyset$, and there is an m-blue ball B_b larger than B_r but smaller than B such that $B_b^2 \cap B_r^2 \neq \emptyset$. Smaller here means that the radius is smaller than or equal to, and larger means that the radius is greater than or equal to. There is a symmetric definition for blue balls which is obtained by exchanging red with blue in the above definition.

For each integer m we want to find a subcollection of m-red balls in B_0 essentially by a covering argument; in our arguments we replace "disjoint" by "disjoint or nested with" and "contained in B^1" by "contained in B^1 and not nested with B." More precisely, we want to find a subcollection $\{B_j\} = \mathcal{R}_m$ of the m-red balls contained in B_0^1 such that

(i) Any two disitinct balls in $\mathcal{R}_m$ are disjoint or one is nested with the other.

(ii) If $B_r \subseteq B_0^1$ is an m-red ball, then there is a ball $B_j \in \mathcal{R}_m$ such that $B_r \subseteq B_j^1$ and B_r is not nested with B_j^1.

(iii) If $B_r \subseteq B_0^1$ is an m-red ball, $B_r \notin \mathcal{R}_m$ and $B_j^1 \subseteq B_r$ for some $B_j \in \mathcal{R}_m$, then B_j is nested with B_r.

In fact, to construct such a collection $\mathcal{R}_m$ we need only add a few words to the argument used to prove the usual covering lemma. First we select B_1 to be a large ball in $\mathcal{R}_m$, i.e., radius of $B_1 = \sup(\text{radius of balls in } \mathcal{R}_m)/2$ will do; if $m < 0$ we choose $B_1 = B_0^1$. Suppose that $B_1, \ldots, B_{j-1}$ have been chosen in such a way that (i)-(iii) above hold. For our next choice we only consider m-red balls in B_0^1 which are not contained in any of the previously selected balls $B_1, \ldots, B_{j-1}$ without being nested with the corresponding ball B_i. (i) and (ii) follow by the construction and Lemma 2 below; (iii) follows by the construction and Lemma 3 below.

Lemma 2. Either the collection $\{B_j\}$ is finite or else the radii of the B_j's tends to 0 as $j \to \infty$.

Lemma 3. Assume B_{r_1} and B_{r_2} are m-red balls such that $B_{r_1}^2 \subseteq B_{r_2}^2$, and B is a ball containing B_{r_2}. If B_{r_1} is nested with B, then B_{r_1} is nested with B_{r_2} or B_{r_2} is nested with B.

To avoid unnecessary distractions we prove these lemmas towards the end of the chapter, and proceed with the definition of ϕ^+ and ϕ^-.

Construction of ϕ^+ and ϕ^-

Let $\mathcal{B}_m$ be a collection of m-blue balls obtained in a similar way as $\mathcal{R}_m$. We now order the family $\mathcal{F}_m = \mathcal{B}_m \cup \mathcal{R}_m$ by decreasing size into a sequence $\{B_{j,m}\}$, say. For each m-red ball $B_{j,m} \in \mathcal{F}_m$ of radius $r_{j,m}$, let $a_{j,m}(x)$ be a Lip_α function that satisfies the following properties:

(i) $0 \le a_{j,m}(x) \le 2$.

(ii) $a_{j,m}(x) = 2$ for x in $B_{j,m}^1$.

(iii) The support of $a_{j,m}$ is contained in $B_{j,m}^2$.

(iv) $|a_{j,m}(x) - a_{j,m}(y)| \le c(d(x,y)/r_{j,m})^\alpha$.

Also for each m-blue ball $B_{j,m} \in \mathcal{F}_m$ we let $a_{j,m}(x)$ be a Lip_α function such that $-a_{j,m}$ satisfies (i)-(iv) above.

Next, for each fixed m we add the functions $a_{j,m}$ as we succesively truncate the partial sums when they fall outside of the range $[-1,1]$. Specifically, if with ψ we denote the function $\psi(t) = (\text{sgn}\, t)\min(|t|,1)$, then we define recursively the functions $A_{j,m}$ and $a_{j,m}^*$ by

$$A_{1,m}(x) = a_{1,m}^*(x) = \psi(a_{1,m}(x)) \quad j = 1,$$

and

$$A_{j,m}(x) = \psi(A_{j-1,m}(x) + a_{j,m}(x)) = A_{j-1,m}(x) + a_{j,m}^*(x), \quad j \ge 2.$$

Observe that whenever $B_{j,m} \in \mathcal{R}_m$ we have $A_{j-1,m}(x) + a_{j,m}(x) \ge -1 + 2 = 1$ for $x \in B_{j,m}^1$. It thus follows that

$$A_{j,m}(x) = 1 \quad \text{on } B_{j,m}^1 \quad \text{when} \quad B_{j,m} \in \mathcal{R}_m$$

and similarly,

$$A_{j,m}(x) = -1 \quad \text{on } B_{j,m}^1 \quad \text{when} \quad B_{j,m} \in \mathcal{B}_m.$$

We also have that $a_{j,m}^* \ge 0$ if $B_{j,m} \in \mathcal{R}_m$ and ≤ 0 if $B_{j,m} \in \mathcal{B}_m$, and

$$a_{m,1}^* = \begin{cases} 1 & \text{on } B_0^1 \text{ when } m < 0 \\ -1 & \text{on } B_0^1 \text{ when } m > 0. \end{cases}$$

With a slight abuse of notation, which is useful in what follows, we write $a_{j,m}^* \in \mathcal{R}_m$ whenever $B_{j,m} \in \mathcal{R}_m$, and similarly for the blue balls.

We now define

$$f_m(x) = \lim_{\ell \to \infty} A_{\ell,m}(x) = \sum_{\mathcal{F}_m} a_{j,m}^*(x)$$

$$= \sum_{\mathcal{R}_m} a_{j,m}^*(x) + \sum_{\mathcal{B}_m} a_{j,m}^*(x) = f_m^+(x) + f_m^-(x),$$

say. The following then holds: The sum defining $f_m(x)$ converges μ-a.e., $|f_m(x)| \leq 1$, and

$$f_m(x) = \begin{cases} 1 & \mu\text{-a.e. on } E_m^+ = \{y : \phi(y) - \phi_{B_0} > \lambda(2m+1)\} \cap B_0 \\ 1 & \mu\text{-a.e. on } E_m^- = \{y : \phi(y) - \phi_{B_0} < \lambda(2m-1)\} \cap B_0. \end{cases}$$

This observation is a consequence of the following lemma and the corresponding blue version.

Lemma 4. Suppose x_0 is a Lebesgue point of the set $\{y : \phi(y) - \phi_{B_0} \geq 0\}$. Then there is an integer $j(x_0)$ such that $x_0 \notin B_{j,m}^2$ for all $B_{j,m} \in \mathcal{B}_m$ with $j \geq j(x_0)$. Moreover, if x_0 is also a Lebesgue point of the set E_m^+, then $j(x_0)$ can be chosen so that $j(x_0) \leq j_0$ for some $B_{j_0,m} \in \mathcal{R}_m$ with $x_0 \in B_{j_0,m}^1$.

This lemma will be proved later on.

Now set

$$f = \left(f_0 + \sum_{m \neq 0} (f_m - a_{m,1}^*) \right) \lambda$$

$$= \left(\sum_{m \geq 0} f_m^+ + \sum_{m < 0} (f_m^+ - a_{m,1}^*) \right) \lambda + \left(\sum_{m \leq 0} f_m^- + \sum_{m > 0} (f_m^- - a_{m,1}^*) \right) \lambda$$

$$= f^+ + f^-,$$

say, and define g by the relation $\phi = f^+ + f^- + g + \phi_{B_0}$. From the above properties we conclude that $\|g\|_\infty \leq 4\lambda$.

Finally we set

$$\phi^+ = f^+ \quad \text{and} \quad \phi^- = f^- + g + \phi_{B_0}.$$

Verification of the Properties of ϕ^+ and ϕ^-

Since the verification can be carried out in a similar fashion for both functions we only discuss ϕ^+ here. First we consider (a) and (c).

Let B be a fixed ball contained in B_0; we may assume that B is much smaller than B_0. Note that

$$\phi^+ = \lambda \sum_m \sum_{\mathcal{R}_m \setminus \{B_0^1\}} a_{j,m}^* \,.$$

We split the balls in $\mathcal{R}_m \setminus \{B_0^1\}$ into three families, to wit:

1. Those $B_{j,m}$'s such that $B_{j,m}^2 \cap B = \emptyset$. Since $a_{j,m}^* = 0$ on B these balls can be disregarded.
2. Those $B_{j,m}$'s with radius $r_{j,m} > r = $ radius of B; we call these the large balls.
3. Those $B_{j,m}$'s contained in B_0^1; we call these the small balls.

Thus we may further write $\phi^+(x) = \phi_L^+(x) + \phi_S^+(x)$, where ϕ_L^+ and ϕ_S^+ are the sums of the $a_{j,m}^*$'s extended to the large and the small balls, respectively.

The following two estimates hold:

$$|\phi_L^+(x) - \phi_L^+(y)| \le c\lambda \quad \text{all} \quad x, y \in B\,, \qquad\qquad (d)$$

and

$$\mu(\{x \in B : \phi_L^+(x) > t\}) \le ce^{-t(\beta_1 - \delta/2)}\mu(B)\,. \qquad\qquad (e)$$

(a) and (c) then follow from these estimates, with β_1 replaced by $\beta_1 - \varepsilon$.

Here are the lemmas needed for this purpose, they are proved at the end of the chapter.

Lemma 5. If $B_1 = B(x, r_1)$ is a red ball nested with a ball $B = B(y, r)$, then $r_1 \le cre^{-2\lambda\gamma_1}$; if it is a blue ball, then $r_1 \le cre^{-2\lambda\gamma_2}$. Also $B_1^3 \subseteq B^3$.

Lemma 6. Suppose $B_j \in \mathcal{R}_m$, $B_i \in \mathcal{R}_{m_1}$ and $B_j \ne B_0^1 \ne B_i$. Then,
(i) If $m \ne m_1$, one of the balls is nested with the other or $B_i^2 \cap B_j^2 = \emptyset$.
(ii) If $m = m_1$, one of the balls is nested with the other or $B_i \cap B_j = \emptyset$.

From Lemmas 5 and 6 we get

Corollary 7. Suppose $\{B_j\}$ is a family of balls in $\bigcup_m \mathcal{R}_m \setminus \{B_0^1\}$ such that $r \le$ radius of $B_j \le 2r$ for all j. Then the balls are pairwise disjoint and $\sum_j \chi_{B_j^2} \le c$.

One of the basic estimates is given by

Lemma 8. If $\{B_j\}$ is a collection of pairwise disjoint balls in $\bigcup_m \mathcal{R}_m \setminus \{B_0^1\}$, every one of which is nested with a ball B, then

$$\sum_j \mu(B_j) \le ce^{-\lambda\beta_1}\mu(B)\,.$$

Lemma 9. Suppose $\{B_{j,m}\} \subset \mathcal{F}_m$ is a family of balls of the same color, none of which is nested with another. Then $\sum_j |a_{j,m}^*(x)| \leq 2$.

Lemma 10. If $r_{j,m}$ denotes the radius of $B_{j,m} \in \mathcal{F}_m$, then for all x, y we have

$$|a_{j,m}^*(x) - a_{j,m}^*(y)| \leq c(d(x,y)/r_{j,m})^\alpha.$$

Assuming for the moment that the above results have been proved, (d) is obtained as follows: If $x, y \in B = B(x,r)$, and if $\mathcal{L}$ denotes the collection of large balls in $\bigcup_m \mathcal{R}_m$, then

$$|\phi_L^+(x) - \phi_L^+(y)| \leq \sum_{\mathcal{L}} |a_{j,m}^*(x) - a_{j,m}^*(y)|$$

$$\leq c \sum_{\mathcal{L}} (d(x,y)/r_{j,m})^\alpha (\chi_{B_{j,m}^2}(x) + \chi_{B_{j,m}^2}(y))$$

$$\leq cd(x,y)^\alpha \sum_{k=1}^\infty (2^k r)^{-\alpha} \sum_{1/2 < \frac{r_{j,m}}{2^k r} < 1} (\chi_{B_{j,m}^2}(x) + \chi_{B_{j,m}^2}(y))$$

$$\leq cd(x,y)^\alpha r^{-\alpha} \leq c.$$

As for (e), we start out by sorting all small red balls into collections which we call $\mathcal{G}_\ell$ so that each collection consists of balls that are not nested. First we order all the small red balls by decreasing size so as not to disturb the ordering in each $\mathcal{F}_m$. To get $\mathcal{G}_1$ we go through the collection of small red balls and we separate a ball for $\mathcal{G}_1$ if and only if it is not nested with those balls that were previously selected for $\mathcal{G}_1$. For the remaining small red balls we repeat the selection procedure and obtain a collection of nonnested balls that we call $\mathcal{G}_2$, and so on. This procedure exhausts all small red balls since the first N such balls must be in $\bigcup_{\ell < N} \mathcal{G}_\ell$. Moreover, the $\mathcal{G}_\ell$'s satisfy the following properties:
1. The balls in $\mathcal{G}_\ell$ are not nested.
2. If $B_{j,m} \in \mathcal{G}_\ell$, $\ell > 1$, there is a ball $B_{i,m_1} \in \mathcal{G}_{\ell-1}$ such that $B_{j,m}$ is nested with B_{i,m_1}.

From Lemma 6 it also follows that
3. Balls in $\mathcal{G}_\ell$ are pairwise disjoint.

Let

$$E_\ell = \bigcup_{B_{j,m} \in \mathcal{G}_\ell} B_{j,m}^3.$$

By property 2 above and Lemma 5 it follows that

$$E_\ell \subseteq E_{\ell-1} \subseteq \ldots \subseteq E_1 \subseteq B^4.$$

Whence, by Lemma 8 we obtain

$$\mu(E_\ell) \leq ce^{-\lambda\beta_1}\mu(E_{\ell-1}) \leq c^{\ell-1}e^{-\lambda(\ell-1)\beta_1}\mu(B). \qquad (f)$$

Since by Lemmas 6 and 8 we get that $\sum_{B_{j,m}\in\mathcal{G}_\ell}|a_{j,m}^*| \leq 2$, from the properties of the E_j's we conclude that

$$\{x \in B : \phi_S^+(x) > 2\ell\lambda\} \subseteq E_{\ell+1}.$$

Thus, if λ is sufficiently large, (e) follows from (f). This concludes the verification of (a) and (c); we pass now to that of (b).

Let $B_1 = B(x,r_1) \subseteq B_2 = B(y,r_2)$ be two balls in B_0. As before we only discuss ϕ^+. First note that from (a) and (c) it follows that

$$\phi^+(x) \geq \phi_{B_2}^+ - c, \quad \text{all} \quad x \in B_2,$$

and since $r_1 \leq cr_2$ we get

$$\phi_{B_2}^+ - \phi_{B_1}^+ \leq c \leq \frac{1}{\gamma_1}(\ln r_2 - \ln r_1) + c.$$

So it remains to show that

$$\phi_{B_2}^+ - \phi_{B_1}^+ \geq -\frac{1}{\gamma_2^{-\delta}}(\ln r_2 - \ln r_1) - c.$$

For this purpose we split the balls in $\bigcup_m \mathcal{R}_m \setminus \{B_1^0\}$ into four classes, to wit:

(1) Balls $B_{j,m}$ such that $B_{j,m}^2 \cap B_1 = \emptyset$. Since the $a_{j,m}^*$'s may only increase $\phi_{B_2}^+$ but they vanish on B_1, these balls can be disregarded.

(2) Balls which are large relative to B_2. If ϕ_L^+ denotes the corresponding partial sum, the estimate

$$|\phi_L^+(x) - \phi_L^+(y)| \leq c, \quad \text{all} \quad x, y \in B_2,$$

is obtained as before. Whence, these partial sums can be neglected.

(3) Balls which are small relative to B_1 and which are contained in B_1^1. If ϕ_S^+ denotes the corresponding partial sum, as above we get the estimate

$$\mu(\{x \in B_1 : \phi_S^+(x) > t\}) \leq ce^{-t(\beta+\delta/2)}\mu(B_1).$$

Thus $(\phi_S^+)_{B_1} \leq c$, and since $\phi_S^+(x) \geq 0$ it follows that $(\phi_S^+)_{B_2} \geq 0$, and consequently, $(\phi_S^+)_{B_2} - (\phi_S^+)_{B_1} \geq -c$.

(4) All the remaining balls, namely those balls of intermediate size, larger than B_1 but smaller than B_2; let ϕ_M^+ denote the corresponding partial sum. We show below that

$$0 \leq \phi_M^+(x) \leq -\frac{1}{\gamma_2^{-\delta}}(\ln r_2 - \ln r_1) - c.$$

From this estimate, and those for the cases (1)-(3), the desired estimate for $\phi_{B_2}^+ - \phi_{B_1}^+$ follows.

To prove the assertion in case (4), we split the intermediate size balls into collections $\mathcal{G}_\ell$, $\ell = 1, 2, \ldots$ of nonnested balls such that each ball in $\mathcal{G}_\ell$, $\ell \geq 2$, is nested with a ball in $\mathcal{G}_{\ell-1}$. First we count how many such nonempty collections can occur. If $r(\ell)$ is the minimum of the radii of the balls in $\mathcal{G}_\ell$, by Lemma 5 we have

$$r(\ell) \geq r(\ell-1) \frac{1}{c} e^{2\gamma_1 \lambda},$$

and since $r_1 \leq r(1)$ and $r(\ell) \leq r_2$ whenever $\mathcal{G}_\ell \neq \emptyset$, we get

$$\left(e^{2\gamma_1 \lambda}/c\right)\ell \leq cr_2/r_1.$$

Thus

$$\ell \leq \frac{1}{(2\gamma_1 \lambda - c)}(\ln r_2 - \ln r_1) - c.$$

From Lemmas 6 and 9 as above we get that $0 \leq \sum_{\mathcal{G}_\ell} a_{j,m}^*(x) \leq 2$. Whence,

$$0 \leq \phi_M^+(x) \leq \frac{1}{(\gamma_1 - (c/2\lambda))}(\ln r_2 - \ln r_1) - c,$$

which finally gives the desired estimate provided that λ is chosen large enough.

To complete the proof of Theorem 1 it only remains to prove Lemmas 2-6 and 8-10; we begin with some preliminary observations.

Note that, by (a) or (b), if two balls intersect and the difference of the median values is large, then one must be much larger than the other. Thus, for instance,

$$|\phi_B - \phi_{B^k}| \leq ck, \quad k = 1, 2, \ldots \tag{g}$$

Whence, if B_b is m-blue and B_r is m-red, and if $B_b^2 \cap B_r^2 = \emptyset$, then one ball is much larger than the other. More precisely, if B_b is the larger ball, by (b) it follows that

$$(\text{radius of } B_b) \geq ce^{2\gamma_2 \lambda}(\text{radius of } B_r).$$

From this remark we conclude the first part of Lemma 5, the second part is then easy.

By Lemma 5 it follows that the balls in the sequence $\{B_j\}$ in Lemma 2 which are roughly of the same size cannot be nested. Therefore these balls must be pairwise disjoint, and since they are all contained in B_0^1 there can only be a finite number of them. Lemma 2 now follows easily. As for Lemma 3, just consider two cases, namely, when the m-blue ball which makes B_1 nested with B_2 is larger or smaller than B_2.

Proof of Lemma 4. The first half follows readily from Lemma 2 and the definitions of Lebesgue point of density and of median value. As for the last part, let $j(x_0)$ be chosen so that $B^2_{j(x_0)}$ is the last m-blue ball in $\mathcal{F}_m$ that contains x_0; if no such ball exists put $j(x_0) = 0$. By the Lebesgue density theorem there is a smaller m-red ball B_r containing it. If $B_r \in \mathcal{R}_m$, then $B_r = B_j$ with $j > j(x_0)$. On the other hand, if $B_r \notin \mathcal{R}_m$, by property (ii) in the definition of $\mathcal{R}_m$ there exists $B_j \in \mathcal{R}_m$ such that $B_r \subseteq B^1_j$, but it is not nested with it. Thus $B_{j(x_0)}$ is larger than B_j and $j > j(x_0)$. ∎

Before we proceed with the proof of Lemma 6 we show that if $B_j \in \mathcal{R}_m$ and $B_j \neq B^1_0$, then

$$(2m+1)\lambda - c_2 < \phi_{B_j} - \phi_{B_0} < (2m+1)\lambda - c_2/2\,,$$

provided that $c_2 > 0$ is large enough. The left-hand side of the above inequality holds by definition. Also, since by Lemma 5 B^1_j cannot be nested by B_j, from property (iii) of $\mathcal{R}_m$ we conclude that B^1_j is not m-red. Whence by (g) above, the right-hand side inequality is also true.

Proof of Lemma 6.   We do (i) first. We may assume that B_i is larger than B_j, and begin by considering the case $m > m_1$. By the above inequality B_i is m-blue. Thus, if $B^2_i \cap B^2_j \neq \emptyset$, the balls are nested. On the other hand, if $m < m_1$, then the ball B^3_i is m-red but it does not satisfy the right-hand side of the above inequality; thus $B^3_j \notin \mathcal{R}_m$. If $B^2_i \cap B^2_j \neq \emptyset$, then $B^1_j \subseteq B^3_i$, and by the property (iii) of $\mathcal{R}_m$ we conclude that B_j is nested with B^3_i. Also, since the involved m-blue ball is much smaller than B^3_i, B_j is nested with B_i. As for (ii), it is just property (i) of $\mathcal{R}_m$. ∎

Proof of Lemma 9.   Fix x. If $B_{j,m}, B_{k,m} \in \mathcal{R}_m$ are not nested, and if $a^*_{j,m} \neq 0$ and $a^*_{k,m} \neq 0$, then $a^*_{i,m}(x) \geq 0$ for all $j \leq i \leq k$. We then conclude that

$$\sum_{i=j}^{k} |a^*_{i,m}(x)| = \sum_{i=j}^{k} a^*_{i,m}(x) = A_{k,m}(x) - A_{j,m}(x) \leq 2\,. \quad ∎$$

Proof of Lemma 10.   Observe that the truncations are also contractions. Thus

$$|A_{j,m}(x) - A_{j,m}(y)| \leq \sum_{i=1}^{m} |a_{j,m}(x) - a_{i,m}(y)|\,,$$

and if $r_{j,m}$ denotes the radius of $B_{j,m}$ and $r = r_{i,m}$, by Corollary 7 we get

$$|a^*_{i,m}(x) - a^*_{i,m}(y)| \leq |A_{i,m}(x) - A_{i,m}(y)| + |A_{i-1,m}(x) - A_{i-1,m}(y)|$$

$$\leq 2\sum_{j=1}^{i}|a_{j,m}(x)-a_{j,m}(y)|$$

$$\leq 2\sum_{j=1}^{i}d(x,y/r_{j,m})^{\alpha}\left(\chi_{B_{j,m}^{2}}(x)+\chi_{B_{j,m}^{2}}(y)\right)$$

$$\leq 2d(x,y)^{\alpha}\sum_{k=1}^{m}(2^{k}r)^{-\alpha}\sum_{r_{j},m\sim2^{k}r}\left(\chi_{B_{j,m}^{2}}(x)+\chi_{B_{j,m}^{2}}(y)\right)$$

$$\leq d(x,y)^{\alpha}\sum_{k=1}^{m}(2^{k}r)^{-\alpha}=c(d(x,y)/r)^{\alpha}.\quad\blacksquare$$

It now only remains to prove Lemma 8; this is the basic estimate in the construction. As a first step we invoke (a) to prove

Lemma 11. If $\{B_{i}\}$ is a collection of pairwise disjoint balls contained in a fixed ball B, then

$$\sum_{i}\mu(B_{i})\leq ce^{-s\beta_{1}}\mu(B)\quad\text{if}\quad\text{for all }i,\phi_{B_{i}}\geq\phi_{B}+s,s>0,$$

and

$$\sum_{i}\mu(B_{i})\leq ce^{-s\beta_{2}}\mu(B),\quad\text{if}\quad\text{for all }i,\phi_{B_{i}}\leq\phi_{B}+s,s>0.$$

Proof. Since both inequalities are proved in a similar fashion we only prove the first. By (a) it readily follows that

$$\sum_{i}\mu(B_{i})\leq\sum_{i}e^{-\beta_{1}\phi_{B_{i}}}\int_{B_{i}}e^{\beta_{1}\phi(x)}d\mu$$

$$\leq 2e^{-\beta_{1}s}e^{-\beta_{1}\phi_{B}}\int_{\cup_{i}B_{i}}e^{\beta\phi(x)}d\mu$$

$$\leq 2e^{-\beta_{1}s}e^{-\beta_{1}\phi_{B}}\int_{B}e^{\beta\phi(x)}d\mu\leq e^{-\beta_{1}s}\mu(B).\quad\blacksquare$$

Proof of Lemma 8. Let m_{0} be an integer such that $(2m_{0}-1)\lambda\leq\phi_{B}-\phi_{B_{0}}\leq(2m_{0}+1)\lambda$. We estimate now different subcollections of $\{B_{j}\}$. First, for balls in $\{B_{j}\}\cap\bigcup_{m>m_{0}}\mathcal{R}_{m}$, the estimate follows from Lemma 11. Next, when $m\leq m_{0}$, we estimate the measure of $\{B_{j}\}\cap\mathcal{R}_{m}$. We do this by first considering all m-blue balls B_{b} contained in B^{3}. By a covering argument we find a pairwise disjoint collection of m-blue balls $\{B_{b,i}\}$, say, such that for every m-ball B_b there is a $B_{b,i}$ such that $B_b^3 \subseteq B_{b,i}^4$. By Lemma 11 it follows that

$$\sum_i \mu(B_{b,i}) \leq ce^{-\beta_2(m-m_0)\lambda}\mu(B_b).$$

Since all balls in $\{B_j\} \cap \mathcal{R}_m$ are nested with B, each ball is contained in some ball $B_b^3 \subseteq B_{b,i}^4$. From Lemma 11 again it follows that

$$\sum_{\{B_j\}\cap\mathcal{R}_m, B_j\subseteq B_{b,i}^4} \mu(B_j) \leq ce^{-\beta_1\lambda}\mu(B_{b,i}).$$

From these two estimates we get

$$\sum_{\{B_j\}\cap\mathcal{R}_m} \mu(B_j) \leq ce^{-\beta_1+(m_0-m)\beta_2)}\mu(B), \quad m \leq m_0.$$

Summing now over all $m \leq m_0$ we conclude that the assertion also holds for $\{B_j\} \cap \bigcup_{m \geq m_0} \mathcal{R}_m$, and we are done. ∎

Limiting Argument for the General Case

So far we have checked all the properties of our functions restricted to a fixed large ball B_0. To remove this assumption we set $B_0 = B(x_0, r)$ and let $r \to \infty$. Specificaly, let av ϕ, av ϕ^+ and av ϕ^- denote the mean or average value of ϕ, ϕ^+ and ϕ^- over a fixed ball contained in B_0 respectively. Then

$$\phi - \text{av}\,\phi = (\phi^+ - \text{av}\,\phi^+) + (\phi^- - \text{av}\,\phi^-) \quad \text{on } B_0,$$

and each of the two summands on the right-hand side above has bounded L^2 norm on any compact set K, independent of B_0, provided that $K \subseteq B_0$. Hence we can find a sequence of functions $\phi^+ - \text{av}\,\phi^+$ which converges weakly on each compact set to some function which we call again $\phi^+ - \text{av}\,\phi^+$; similarly for $\phi^- - \text{av}\,\phi^-$. It now follows that $\phi(x) = \phi^+(x) + \phi^-(x)$, that ϕ^+ and ϕ^- satisfy conditions (a), (b) and (c), and that $w_1 = e^{\phi^+}$ and $w_2 = e^{\phi^-}$ satisfy the desired properties with indices $\beta_i - \delta$ and $\gamma_i - \delta$, $i = 1, 2$, respectively, for all balls B. This completes the proof.

Sources and Remarks. If μ is the Lebesgue measure, the decomposition of weights $w \in A_p(\mu)$ into products $w = w_1 w_2^{1-p}$, with w_1, w_2 in $A_p(\mu)$, is due to P. Jones [1980]. Interesting applications of this result are discussed in that paper and in the paper of D. Kurtz and R. Wheeden [1979].

CHAPTER

III

Sharp Maximal Functions

In this chapter we study the main properties of two different versions of the so-called sharp maximal function. Assume that X, d are as in Chapter I, and that ν is a weighted measure with respect to μ with weight w. In addition, suppose there is a class $\mathcal{P}$ of scalar functions defined on X that satisfies the following three properties:

(i) $\mathcal{P} = \{p\}$ is a finite-dimensional linear space of measurable functions on X.

(ii) There are numbers $0 < s_1, s_2 < 1$ such that for all balls B in X and p in $\mathcal{P}$

$$\mu(\{x \in B : |p(x)| > s_1 \sup_{y \in B} |p(y)|\}) \geq s_2 \mu(B).$$

(iii) $\mathcal{P}$ is closed in the topology of "uniform convergence over compact sets."

The sharp maximal functions of interest to us are defined as follows. If $r > 0$ and $f \in L^r_{\text{loc}}(\mu)$ we put

$$M^{\sharp, \mathcal{P}}_{r,\nu} f(x) = \sup_{B \supset \{x\}} \inf_{p \in \mathcal{P}} \frac{\mu(B)}{\nu(B)} \left(\frac{1}{\mu(B)} \int_B |f - p|^r d\mu \right)^{1/r},$$

and if f is merely measurable and $0 < s < 1$ we let the local sharp maximal function $M^{\sharp, \mathcal{P}}_{0,\nu,s} f(x)$ equal

$$\sup_{B \supset \{x\}} \inf_{p \in \mathcal{P}} \frac{\mu(B)}{\nu(B)} \inf\{A > 0 : \mu(\{y \in B : |f(y) - p(y)| > A\}) \leq s\mu(B)\}.$$

When $\nu = \mu$ we drop the subscript ν in the notation. Likewise, when $\mathcal{P}$ is the collection of constants, we drop the superscript $\mathcal{P}$ in the notation; these conventions apply to all other concepts introduced in this chapter as well.

We study in detail, keeping the applications we make of these concepts in later chapters in mind, three questions, namely, how do the sharp maximal functions behave when: (1) We vary the class $\mathcal{P}$, (2) We consider different values of r and s, and, (3) We have $\nu \neq \mu$.

We begin by introducing another maximal function which is useful in the sequel. Given $0 < s < 1$ and a measurable function f, let

$$M_{0,\mu,s}f(x) = \sup_{B \supset \{x\}} \left(\inf\{A > 0 : \mu(\{y \in B : |f(y)| > A\}) \leq s\mu(B)\}\right).$$

We then have

Lemma 1. Assume μ is a doubling measure, and let Φ be any function on $[0,\infty)$ such that $0 < \Phi(1) < \infty$, and $\Phi(t) = \int_0^t \varphi(u)\,du$ for some nonnegative measurable function φ on $(0,\infty)$ that satifies the growth condition $\varphi(2u) \leq k\varphi(u)$ for all $u > 0$.

We then have

$$\mu(\{x \in X : M_{0,\mu,s}f(x) > \lambda\}) \leq c\mu(\{x \in X : |f(x)| > \lambda/2\})$$

and consequently

$$\int_X \Phi(M_{0,\mu,s}f)\,d\mu \leq c \int_X \Phi(|f|)\,d\mu.$$

Furthermore, if ν is a weighted measure with respect to μ with weight $w \in A_\infty(\mu)$,

$$\int_X \Phi(M_{0,\mu,s}f)\,d\nu \leq c \int_X \Phi(|f|)\,d\nu.$$

Proof. Let $M_{0,\mu,s}^n f(x)$ denote the local maximal function defined as $M_{0,\mu,s}f(x)$ but restricted to balls with radius less than or equal to n. Any continuity assertion proved for $M_{0,\mu,s}^n f$ with constant c independent of n implies the corresponding result for $M_{0,\mu,s}f$ by a limiting argument.

Let then $x \in \{M_{0,\mu,s}^n f > \lambda\}$. Then there is a ball B of radius less than or equal to n containing x such that

$$\mu(\{y \in B : |f(y)| > M_{0,\mu,s}^n f(x)/2\}) > s\mu(B). \tag{1}$$

By the covering lemma, Lemma 1 in Chapter I, there is a countable collection $\{B_j\}$ of these balls so that $\{M_{0,\mu,s}^n f > \lambda\} \subseteq \bigcup_j B_j^1$. Thus, since μ is doubling, there is a constant c, say, independent of n such that

$$\mu(\{M_{0,\mu,s}^n f > \lambda\}) \leq c \sum_j \mu(B_j) \leq cs^{-1} \sum_j \mu(\{y \in B_j : |f(y)| > \lambda/2\})$$

$$\leq cs^{-1}\mu(\{|f| > \lambda/2\}).$$

The first conclusion of the theorem follows, then, by letting n tend to ∞, and the second follows from the first and the identity

$$\int_X \Phi(|f|)\,d\mu = \int_0^\infty \mu(\{|f| > \lambda\})\varphi(\lambda)\,d\lambda.$$

Furthermore, by condition (6) in Chapter I, (1) above implies that for some $p > 1$, also

$$\nu(\{y \in B : |f(y)| > M^n_{0,\mu,s}f(x)/2\}) \ge cs^p\nu(B).$$

This is essentially (1), but with μ replaced by ν, which is also doubling. Thus, the argument given above also applies to ν, and we are done. ∎

We return now to the consideration of the sharp maximal functions. Some relations between these functions are readily obtained. For instance, by Chebychev's inequality, it follows that

$$M^{\sharp,\mathcal{P}}_{0,\nu,s}f(x) \le M^{\sharp,\mathcal{P}}_{r,\nu}f(x)/s^{1/r} \quad \text{all } 0 < s < 1, r > 0.$$

We need some definitions. Let $A^{\mathcal{P}}_r(f,B)$ denote the estimate of the oscillation, modulo $\mathcal{P}$, of f over B given when $r > 0$ by

$$A^{\mathcal{P}}_r(f,B) = \inf_{p \in \mathcal{P}} \left(\frac{1}{\mu(B)} \int_B |f - p|^r d\mu \right)^{1/r},$$

and when $r = 0$ by

$$A^{\mathcal{P}}_{0,s}(f,B) = \inf_{p \in \mathcal{P}} \inf\{A > 0 : \mu(\{y \in B : |f(y) - p(y)| > A\}) < s\mu(B)\}.$$

We remark that when f is real-valued and $\mathcal{P} = \{\text{constants}\}$, a good choice for the constant in the definition of $A_r(f,B)$ is the median value f_B of f over B. When $f = \Re f + i\Im f$ is complex-valued, we define $f_B = (\Re f)_B + i(\Im f)_B$, and note that also this is a good choice for the constant.

It turns out that f_B is also a good choice for the constant in the definition of $A_{0,s}(f,B)$. More precisely, given $x \in X$ and a ball B that contains x, we have

$$\mu(\{y \in B : |f(y) - f_B| > 2M^{\sharp}_{0,s}f(x)\}) < s\mu(B).$$

When $\mathcal{P}$ is more general than the constants, we must replace f_B by an appropriate $p_B \in \mathcal{P}$. In fact, we define the functions $p_B(f)$ and $p_{B,s}(f)$ in $\mathcal{P}$, by the relations

$$A^{\mathcal{P}}_r(f,B) = \left(\frac{1}{\mu(B)} \int_B |f - p_B(f)|^r d\mu \right)^{1/r},$$

and

$$\mu(\{y \in B : |f(y) - p_{B,s}(f)(y)| > A^{\mathcal{P}}_{0,s}(f,B)\}) \le s\mu(B),$$

respectively. The existence of these functions is proved by a limiting argument.
The following estimates are the basic results in this chapter.

Theorem 2. Suppose μ is a doubling measure. Then there exists a constant $0 < s_1 < 1$ with the following property: Given $0 < s \le s_1$, there are positive constants c, c_1 such that for all balls B,

$$\mu(\{y \in B : |f(y) - p_{B,s}(f)(y)| > \lambda, M_{0,s}^{\sharp,\mathcal{P}} f(y) < \alpha\}) \le c e^{-c_1 \lambda / \alpha} \mu(B),$$

for all $\alpha, \lambda > 0$.

Theorem 3. There is a constant $0 < s_1 < 1$ with the following property: If $0 < s \le s_1$ and f is measurable, then

$$\|f - p_f\|_{L^p(\mu)} \le c \|M_{0,s}^{\sharp,\mathcal{P}} f\|_{L^p(\mu)} \quad 0 < p < \infty,$$

where $p_f \in \mathcal{P}$ depends on f.

Proof of Theorem 2. Fix a ball B. For simplicity we consider the case $\mathcal{P} = \{\text{constants}\}$; the argument in the general case follows along similar lines.

We begin by observing that if $M_{0,s}^{\sharp} f(y) < \alpha$ and $0 < s$ is sufficiently small, then

$$|f_{B(y,R)} - f_{B(y,r)}| \le 4\alpha, \quad \text{provided } r \le R \le 2r. \tag{2}$$

Indeed, note that

$$\mu(\{x \in B(y,R) : |f(x) - f_{B(y,R)}| > 2\alpha\}) \le s\mu(B(y,R)) \le sc\mu(B(y,r)).$$

Consequently, if (2) does not hold, we must have $|f(x) - f_{B(y,R)}| > 2\alpha$ for those $x \in B(y,R)$ for which $|f(x) - f_{B(y,R)}| \le 2\alpha$. On the other hand,

$$\mu(\{x \in B(y,r) : |f(x) - f_{B(y,r)}| \le \alpha\}) > (1 - s)\mu(B(y,r)).$$

Thus, if (2) does not hold it follows that $(1 - s) \le cs$, or $s > 1/(1 + c)$ which leads to a contradiction when $s \le s_1$ is suuficiently small; clearly $s_1 \le 1/(1 + c)$ will do.

A similar argument gives the following estimate: Let $r = $ radius of B and suppose that $r' \le r \le 2r'$, and $M_{0,s}^{\sharp} f(y) < \alpha$. Then there is a constant c such that

$$|f_{B(y,r')} - f_B| \le c\alpha. \tag{3}$$

If c is the constant in (3), let $c_2 = \max(4, c)$. The proof of the theorem proceeds as follows: We may as well assume that $\lambda / c_2 \alpha > 1$, for otherwise there is nothing to prove. Since almost every point y of $E = \{y \in B : |f(y) - f_B| > \lambda, M_{0,s}^{\sharp} f(y) < \alpha\}$ is a density point, there is a ball B_y centered at y such that

$$|f_{B_y} - f_B| > \lambda. \tag{4}$$

Thus combining (3) and (4) we can find a ball $B(y, r_y)$ for y μ-a.e. in E such that

$$\lambda < |f_{B(y,r_y)} - f_B| < c_2\alpha + \lambda.$$

By a covering argument there is a countable pairwise disjoint subcollection $\mathcal{F}_\infty = \{B(y_{j,1}, r_{j,1})\}$ of these balls such that except for a null subset of E we have,

$$E \subseteq \tilde{E}_1 = \bigcup_{B_j \in \mathcal{F}_1} B_j^1, \quad B_j^1 \subseteq B^1 \quad \text{all } j.$$

These are the so-called first step balls. Let $E_1 = \bigcup_{B_j \in \mathcal{F}_1} B_j$ denote the union of the first step balls. Since $y_{j,1} \in E$, from (2) it readily follows that $|f_{B(y_{j,1}, r_{j,1})} - f_{B(y_{j,1}, 2r_{j,1})}| \leq c_2\alpha$, and from (3) that $|f_{B(y_{j,1}, r)} - f_B| \leq c_2\alpha$. Thus

$$|f_{B(y_{j,1}, 2r_{j,1})} - f_B| \leq c_2\alpha + c_2\alpha \leq 2c_2\alpha. \tag{5}$$

Whence combining estimates (4) and (5) we see there are balls $B(y_{j,1}, r')$ with $2r_{j,1} \leq r' \leq r$ so that

$$\lambda - 2c_2\alpha \leq |f_{B(y_{j,1}, r')} - f_B| \leq \lambda. \tag{6}$$

Let $\mathcal{F}_2 = \{B(y_{i,2}, r_{i,2})\}$ be a countable pairwise disjoint subcollection of balls satisfying (6) such that

$$E_1 \subseteq \bigcup_{B_i \in \mathcal{F}_2} B_i^1 \quad \mu\text{-a.e.},$$

and let $E_2 = \bigcup_{B_i \in \mathcal{F}_2} B_i$. Note that for each index i there are indices j, j' such that $y_{j,1} = y_{i,2}$ and $B^1(y_{j',1}, r_{j',1}) \subseteq B(y_{i,2}, r_{i,2})$. We repeat this argument k times obtaining in this fashion kth step balls $\mathcal{F}_k = \{B(y_{j,k}, r_{j,k})\}$ and sets $E_k = \bigcup_{B_j \in \mathcal{F}_k} B_j$, $\tilde{E}_k = \bigcup_{B_j \in \mathcal{F}_k} B_j^1$ so that

$$\lambda - c_2(k+1)\alpha \leq |f_{B(y_{j,k}, r_{j,k})} - f_B| \leq \lambda - c_2(k-1)\alpha.$$

Observe that if a ball $B_1 \in \mathcal{F}_\ell$, $1 \leq \ell < k$, then its center is the center of a ball chosen is a previous step, and furthermore there is a ball B_2 in $\mathcal{F}_k$ such that $B_1^1 \subseteq B^2$.

This process can be repeated until $\lambda - c_2(k+1)\alpha \sim 0$, i.e., until $k = k_0 \sim \lambda/c_2\alpha - 1$; at that point we assign to each ball the ball B^1 and stop.

Clearly $E \subseteq E_1 \subseteq \ldots \subseteq E_{k_0} = B^1$. We estimate now the measure of the E_k's. Fix B_0 in $\mathcal{F}_j$ and let $\{B_{i,j-1}\}_i$ be the collection of those pairwise disjoint balls in $\mathcal{F}_{j-1}$ such that $B_{i,j-1}^1 \subseteq B_0$; we then have

$$\mu\left(\bigcup_i B_{i,j-1}^1\right) \leq c\sum_i \mu(B_{i,j-1}). \tag{7}$$

Since from the above properties it follows that

$$\mu(\{x \in B_{i,j-1} : |f(x) - f_{B_{i,j-1}}| > 2\alpha\}) \le s\mu(B_{i,j-1}),$$

we also have

$$\mu(B_{i,j-1}) \le \frac{1}{1-s}\mu(\{x \in B_{i,j-1} : |f(x) - f_{B_{i,j-1}}| \le 2\alpha\}). \tag{8}$$

Next we claim that if $|f(x) - f_{Bi,j-1}| \le 2\alpha$, then $|f(x) - f_{B_0}| > 2\alpha$. Indeed, since $f(y) - f_{B_0} = (f(y) - f_{B_{i,j-1}}) + (f_{B_{i,j-1}} - f_B) + (f_B - f_{B_0})$, it follows that

$$|f(y) - f_{B_0}| \ge -2\alpha + \lambda - c_2 k\alpha - (\lambda - c_2(k+1)\alpha) \ge c_2\alpha > 2\alpha,$$

provided that $c_2 \ge 4$. Thus,

$$\sum_i \mu(\{x \in B_{i,j-1} : |f(x) - f_{B_{i,j-1}}| \le 2\alpha\})$$

$$\le \mu(\{x \in B_0 : |f(x) - f_{B_0}| > 2\alpha\}) \le s\mu(B_0). \tag{9}$$

Whence, combining (7), (8), and (9), it readily follows that $\mu(\bigcup_i B^1_{i,j-1}) \le (cs/(1-s))\mu(B_0)$. Thus, if $\mathcal{F}^0_{j-1} = \{B_{i,j-1} \in \mathcal{F}_{j-1} : B_{i,j-1} \subseteq B_0\}$,

$$\mu(E_{j-1}) \le \sum_{B_0 \in \mathcal{F}_j} \sum_{B_{i,j-1} \in \mathcal{F}^0_{j-1}} \mu(B^1_{i,j-1})$$

$$\le \frac{cs}{1-s} \sum_{B_0 \in \mathcal{F}_j} \mu(B_0) \le \frac{cs}{1-s}\mu(\tilde{E}_j).$$

Now, provided that s is small enough, we have $cs/(1-s) = \beta < 1$, and consequently,

$$\mu(E) \le \beta^{k_0}\mu(B^1) \le c\beta^{k_0}\mu(B).$$

Pick now c_1 so that $\beta^{1/c_2} \sim e^{-c_1}$. Our last estimate can then be rewritten

$$\mu(E) \le ce^{-c_1\lambda\alpha}\mu(B),$$

and we are done. ∎

Proof of Theorem 3. For simplicity we prove the assertion for the case $\mathcal{P} = \{\text{constants}\}$. It suffices to show that if β is a sufficiently large positive constant, we have

$$\mu(\{|f(x) - c_f| > \lambda\}) \le c\sum_{k=0}^{\infty}(c_1 e^{-c\beta})^k \mu(\{M^\sharp_{0,s}f(x) > \lambda/2^k\beta\}), \quad \lambda > 0.$$

First we determine the constant c_f. If X is compact, there is $r_0 > 0$ such that $B = B(x,r_0) = X$ for all $x \in X$; set then $c_f = f_B$. Otherwise, if X is not compact, let $x \in X$ and $r > 1$. Then $\mu(B(x,r)) \geq c(x)r^{\varepsilon_1}$ for some $\varepsilon_1 > 0$, and by an argument similar to the proof of Theorem 2 we have

$$|f_{B(x,2r)} - f_{B(x,r)}| \leq \inf_{y \in B(x,r)} M^{\sharp}_{0,s}f(y)$$

$$\leq c\left(\|M^{\sharp}_{0,s}f\|^p_{L^p(\mu)}/\mu(B(x,r))\right)^{1/p} \leq c_{f,x}r^{-\varepsilon_1/p}.$$

It thus follows that $f_{B(x,2^k)}$ converges to a constant $c_f(x)$ as k tends to ∞; since from the above estimate it also follows that $c_f(x)$ is independent of x, this is our choice for c_f.

We estimate the measure of the set $E = \{y : |f(y) - c_f| > \lambda\}$ by constructing a telescoping sequence of collections $\mathcal{F}_k$ of balls which differs from the previous constructions in that there is a remainder set F_k. More precisely, let $\mathcal{F}_k$ and F_k, $k = 1,2\dots$, satisfy the following three properties:

1. The balls in each of the $\mathcal{F}_k$'s are pairwise disjoint.
2. For each ball $B_i \in \mathcal{F}_{k-1}$, $B_i \not\subseteq F_k$, there is a ball $B_j \in \mathcal{F}_k$ such that $B_i^1 \subseteq B_j^1$.
3. Although there may be infinitely many collections $\mathcal{F}_k$, each ball $B_i \in \mathcal{F}_k$ is contained in a finite telescoping chain of balls

$$B_i^1 \subseteq B^1_{i,k+1}, \quad B_{i,k'} \subseteq B^1_{i,k+1},$$

where $B_{i,k'} \in \mathcal{F}_{k'}$, $B_{i,k_0} \subseteq F_{k_0}$, and $k_0 = k_0(B_i)$ depends on B_i.

As usual, let $E_k = \bigcup_{B_i \in \mathcal{F}_k} B_i$, and $\tilde{E}_k = \bigcup_{B_i \in \mathcal{F}_k} B_i^1$.

Now, suppose that this construction can be carried out for each B_j in $\mathcal{F}_k$ so that $E \subseteq E_1$ and if $\mathcal{F}^j_{k-1} = \{B_i \in \mathcal{F}_{k-1} : B_I \subseteq B_j, B_i \not\subseteq F_{k-1}\}$, we have

$$\mu\left(\bigcup_{B_i \in \mathcal{F}^j_{k-1}} B_i\right) < \gamma\mu(B_j). \tag{10}$$

We claim then that the following estimate holds:

$$\mu(E) \leq \sum_{k=1}^{\infty} \gamma^{k-1}\mu(F_k). \tag{11}$$

Indeed, for each $B_j \in \mathcal{F}_k$ let the "stopping time" $t(B_j)$ be the smallest integer k_0 such that there is a chain $B^1_{j,k} \subseteq B^1_{j,k+1} \subseteq \dots \subseteq B^1_{j,k_0} \subseteq F_{k_0}$, where $B_{j,k} \in \mathcal{F}_k$ for each $k \leq k' \leq k_0$.

Now, for each $m \geq 2$, put $\mathcal{F}_{k,m} = \{B_j \in \mathcal{F}_k : t(B_j) = m\}$, and note that we can use (10) to get

$$\mu\left(\bigcup_{B_j \in \mathcal{F}_{1,m}} B_j\right) \leq \gamma\mu\left(\bigcup_{B_j \in \mathcal{F}_{2,m}} B_j\right) \leq \gamma^{m-2}\mu\left(\bigcup_{B_j \in \mathcal{F}_{m,m}} B_j\right)$$

$$\leq \gamma^{m-2}\mu(F_m).$$

Whence summing over m it follows that

$$\mu(E) \leq \mu\left(\bigcup_{B_j \in \mathcal{F}_1} B_j^1\right) \leq c\mu\left(\bigcup_{B_j \in \mathcal{F}_1} B_j\right) \leq c\sum_{m=2}^{\infty} \gamma^{m-2}\mu(F_m),$$

completing the proof of (11).

It thus remains to construct the collections $\mathcal{F}_k$ and the remainder sets F_k. Let $F_k = \{M_{0,s}^{\sharp}f(y) > \lambda/2^k\beta\}$, where β is a large positive constant yet to be chosen. When X is not compact we may have that for an integer k_0, $F_k = X$ for all $k \geq k_0$ and $F_k \neq X$ for $k < k_0$. If this is not the case, i.e., if $F_k \neq X$ for all k we set $k_0 = \infty$.

Let $\{B(x, r_x)\}_{x \in E}$ be a collection of balls such that $|f_{B(x,r_x)} - c_f| > \lambda$; by a Lebesgue density argument such balls exist for x μ-a.e. on E. By a covering argument we can find a pairwise disjoint subcollection $\{B_j\} = \mathcal{F}_1$ of these balls such that $E \subseteq \bigcup_{B_j \in \mathcal{F}_1} B_j^1$. Now suppose that $\mathcal{F}_{k-1}$ has been constructed and that $k < k_0$. For each $B_j \in \mathcal{F}_{k-1}$ which is not contained in $\mathcal{F}_{k-1}$ we have $\inf_{x \in B_J} M_{0,s}^{\sharp}f(x) \leq \lambda/2^{k-1}\beta$. We also assume that every $B_j \in \mathcal{F}_{k-1}$ satisfies

$$|f_{B_j} - c_f| > 2^{k-2}\lambda. \tag{12}$$

As in the proof of Theorem 2 it is now possible to find for each B_i in $\mathcal{F}_{k-1}$ a ball B_i' such that

$$\lambda/2^k < |f_{B_i'} - c_f| < \lambda(1 + (c\beta))/2^k,$$

provided that B_i is not contained in F_{k-1}. By a covering argument we get a pairwise disjoint subcollection $\mathcal{F}_k$, say, of those balls B_i' such that each B_i' is contained in B_j^1, $B_j \in \mathcal{F}_k$. It then follows that if $B_i \in \mathcal{F}_{k-1}$, either $B_i \subseteq F_{k-1}$ or $B_i \subseteq B_j^1$ for some B_j in $\mathcal{F}_k$. Furthermore, the estimate (12) holds with $k - 2$ there replaced by $k - 1$. Whence from (12), (13), and Theorem 2 we conclude that for each $B_j \in \mathcal{F}_k$ we have

$$\mu\left(\bigcup_{B_i \in \mathcal{F}_{k-1}^j} B_i\right) \leq ce^{-c\beta}\mu(B_j).$$

From this estimate we get (10) and the desired estimate for $\mu(E)$. ∎

We continue now with the consideration of the second item in our list at the beginnig of the chapter, i.e., the relations for different values of r and s. Combining the comments following Lemma 1 with Hölder's inequality it readily follows that

$$M_{0,s}^{\sharp,\mathcal{P}}f(x) \leq cM_r^{\sharp,\mathcal{P}}f(x) \leq c_1 M_{r_1}^{\sharp,\mathcal{P}}f(x), \quad 0 < r < r_1, \ 0 < s < 1.$$

It is interesting to point out that the local sharp maximal function controls, in the norm sense, the sharp maximal function. This result follows from the "factorization" inequality of $M_r^{\sharp,\mathcal{P}}$ in terms of M_μ, which increases functions, and $M_{0,s}^{\sharp,\mathcal{P}}$, which decreases functions, proved in Proposition 4 below.

First an observation: The argument given in the proof of Theorem 3 also holds locally. More precisely, there is a constant $0 < s_1 < 1$ such that for each ball B and all measurable functions f, there is a $p \in \mathcal{P}$ such that

$$\mu(\{x \in B : |f(x) - p(x)| > \lambda\})$$

$$\leq c \sum_{k=0}^{\infty} (e^{-c\beta})^k \mu(\{x \in B : M_{0,s}^{\sharp,\mathcal{P}} f(x) > \lambda/2^k \beta\})$$

for all $\lambda > 0$ provided β is a sufficiently large positive constant and $0 < s \leq s_1$.

Proposition 4. There is a constant $0 < s_1 < 1$ such that if $0 < s \leq s_1$ and $r > 0$,

$$M_r^{\sharp,\mathcal{P}} f(x) \leq c \left(M_\mu \left(M_{0,s}^{\sharp,\mathcal{P}} f \right)^r \right)(x)^{1/r}.$$

Proof. Let $x \in X$ and let B be a ball containing x. Multiply through the above local distribution function inequality by $r\lambda^{r-1}$ and integrate over $(0\,\infty)$ to obtain

$$\int_B |f(x) - p(x)|^r d\mu(x) \leq c_1 \sum_{k=0}^{\infty} e^{-c\beta} (2^k \beta)^r \int_B M_{0,s}^{\sharp,\mathcal{P}} f(x)^r d\mu$$

$$\leq c_1 \int_B M_{0,s}^{\sharp,\mathcal{P}} f(x)^r d\mu,$$

provided β is large enough. The conclusion follows now by dividing this inequality through by $\mu(B)$, and then taking the sup over all balls B that contain x. ∎

Next we consider the dependence of $M_r^{\sharp,\mathcal{P}} f$ on the class $\mathcal{P}$, that is, question (1). Clearly if the classes $\mathcal{P}_0 \subseteq \mathcal{P}$ satisfy (i)-(iii) above, then

$$M_r^{\sharp,\mathcal{P}} f(x) \leq M_r^{\sharp,\mathcal{P}_0} f(x).$$

As for the opposite inequality we have

Lemma 5. Suppose μ is a doubling measure that satisfies a reverse doubling condition, and that $\mathcal{P}_0 \subseteq \mathcal{P}$ satisfy properties (i)-(iii) above. Further assume that there is a constant $\gamma > 0$ with the following property: To each $p \in \mathcal{P}$ there corresponds $p_0 \in \mathcal{P}_0$ such that

$$\sup_{y \in B_1} |p(y) - p_0(y)| \leq c(\mu(B_1)/\mu(B_2))^\gamma \sup_{y \in B_2} |p(y)|, \quad \text{all} \quad B_1 \subseteq B_2.$$

Then, given $f \in L_{\text{loc}}^r(\mu)$, there exists $p_f \in \mathcal{P}$ such that

$$M_r^{\natural, \mathcal{P}_0}(f - p_f)(x) \le c M_r^{\natural, \mathcal{P}} f(x).$$

Our first task is to find the function p_f. If X is compact, then X is a ball B and we put $p_f = p_B(f)$, where $p_B(f)$ is the function introduced in the definition of $A_r^{\mathcal{P}}(f, B)$. On the other hand, if X is not compact, let $x \in X$. We would like to take p_f as the limit of the functions $p_{B(x,r)}(f)$ as r tends to ∞, but this limit does not necessarily exist. Since p_f is determined modulo functions in $\mathcal{P}_0$, we begin by considering the function $p_k^0 \in \mathcal{P}_0$ such that

$$\sup_{B(x,2^{\epsilon k})} |p_{B(x,2^k)}(f) - p_{B(x,2^{k-1})}(f) - p_k^0|$$

$$\le \left(\mu(B(x,2^{\epsilon k}))/\mu(B(x,2^{k-1}))\right)^{\gamma} \sup_{B(x,2^{k-1})} |p_{B(x,2^k)}(f) - p_{B(x,2^{k-1})}(f)|$$

$$\le c 2^{-c_1(1-\epsilon)\gamma k} \left(A_r^{\mathcal{P}}(f, B(x,2^k)) + A_r^{\mathcal{P}}(f, B(x,2^{k-1}))\right). \tag{13}$$

In the last inequality we used the reverse doubling property of μ, property (ii), and the definition of $A_r^{\mathcal{P}}$. Once this is done, put

$$p_f = p_{B(x,1)}(f) + \sum_{k=1}^{\infty} \left(p_{B(x,2^k)}(f) - p_{B(x,2^{k-1})}(f) - p_k^0\right).$$

Thus, if $M_0^{\natural, \mathcal{P}} f$ is not identically infinite, the sum defining p_f converges uniformly on each compact set.

Fix now a ball B and observe that it suffices to show that

$$A_r^{\mathcal{P}_0}(f - p_f, B) \le c \sup_{B_1 \supseteq B} A_r^{\mathcal{P}}(f, B_1). \tag{14}$$

First we show that there is a function $p_0 \in \mathcal{P}_0$ such that

$$\sup_B |p_B(f) - p_0| \le c \sup_{B_1 \supseteq B} A_r^{\mathcal{P}}(f, B_1);$$

this clearly implies that (14) holds. Let k_0 be so large that $B \subseteq B(x, 2^{\epsilon k_0})$. Then p_f is equal to

$$\sum_{k=k_0+1}^{\infty} \left(p_{B(x,2^k)}(f) - p_{B(x,2^{k-1})}(f) - p_k(f)\right) + p_{B(x,2^k)}(f) + \sum_{k=1}^{k_0} p_k^0.$$

By (13) the first sum above is bounded on B by $c \sup_{B_1 \supseteq B} A_r^{\mathcal{P}}(f, B_1)$; also the last sum above is in $\mathcal{P}_0$. Suppose now that $B = B(x_1, r)$ and let k_1 be an integer such that $B(x_1, 2^k r) \subseteq B(x, 2^{k_0}) \subseteq B(x_1, 2^{k_1+c})$. Then,

$$\sup_B |p_{B(x_1,2^{k_1}r)}(f) - p_{B(x,2^{k_0})}(f)| \le c \sup_{B_1 \supseteq B} A_r^{\mathcal{P}}(f, B_1),$$

and it follows that

$$p_B(f) = p_{B(x_1,2^{k_1}r)}(f) + \sum_{k=1}^{k_1} p_{B(x_1,2^k r)}(f) - p_{B(x_1,2^{k-1}r)}(f).$$

Consequently, we can find a function $p_k^1 \in \mathcal{P}_0$ such that

$$\sup_B \left| p_{B(x_1,2^k r)}(f) - p_{B(x_1,2^{k-1}r)}(f) - p_k^1 \right|$$

$$\leq c \left(\mu(B)/\mu(B(x_1,2^{k-1}r)) \right)^\gamma \sup_{B(x_1,2^{k-1}r)} \left| p_{B(x_1,2^k r)}(f) - p_{B(x_1,2^{k-1}r)}(f) \right|$$

$$\leq c 2^{-c\gamma} \sup_{B_1 \supseteq B} A_r^{\mathcal{P}}(f,B_1).$$

It then follows that

$$\sup_B \left| p_f - p_B(f) - \sum_{k=1}^{k_0} p_k^0 + \sum_{k=1}^{k_1} p_k^1 \right| \leq c \sup_{B_1 \supseteq B} A_r^{\mathcal{P}}(f,B_1),$$

where $\sum_{k=1}^{k_0} p_k^0 - \sum_{k=1}^{k_1} p_k^1 \in \mathcal{P}_0$. ∎

Finally, assume that ν is a weighted measure with respect to μ with weight w, and consider question (3) above. Recall that

$$M_{r,\nu}^{\sharp,\mathcal{P}} f(x) = \sup_{x \in B} \frac{\mu(B)}{\nu(B)} A_r^{\mathcal{P}}(f,B).$$

Also, since $M_r^{\sharp,\mathcal{P}} f(y) \geq A_r^{\mathcal{P}}(f,B)$ for all $y \in B$, it follows that

$$M_{r,\nu}^{\sharp,\mathcal{P}} f(x) \leq \sup_{x \in B} \frac{1}{\nu(B)} \int_B \left(M_r^{\sharp,\mathcal{P}} f(y)/w(y) \right) d\nu(y)$$

$$= M_\nu \left((M_r^{\sharp,\mathcal{P}} f)/w \right)(x). \tag{15}$$

Now, if $w \in A_\infty(\mu)$, for each $s > 0$ there is a constant c_s such that for each ball B,

$$\mu(\{y \in B : w(y) \leq \nu(B)/\mu(B) \leq c_s w(y)\}) \geq (1-s)\mu(B),$$

and

$$\nu(\{y \in B : c_s^{-1} w(y) \leq \nu(B)/\mu(B) \leq c_s w(y)\}) \geq (1-s)\nu(B).$$

From these inequalities we can sharpen (15) to

$$M_{r,\nu}^{\sharp,\mathcal{P}} f(x) \leq c_s M_{0,\nu,s}((M^{\sharp,\mathcal{P}} f)/w)(x). \tag{16}$$

Similarly, we readily conclude that

$$M_r^{\sharp,\mathcal{P}} f(x) \le c_s M_\mu (w M_{r,\nu}^{\sharp,\mathcal{P}}) f(x), \qquad (17)$$

and, provided that $w \in A_\infty(\mu)$, also that

$$M_r^{\sharp,\mathcal{P}} f(x) \le c_s M_{0,\mu,s}(w M_{r,\nu}^{\sharp,\mathcal{P}} f)(x). \qquad (18)$$

Using the norm inequalities for maximal operators, and with the notation $d\nu_{1-p}(x) = w(x)^{1-p} d\mu(x)$, these inequalities imply that

$$\|M_{r,\nu}^{\sharp,\mathcal{P}} f\|_{L^p(\nu)} \le \|M_r^{\sharp,\mathcal{P}} f\|_{L^p(\nu_{1-p})},$$

and if $d\nu_p(x) = w(x)^p d\mu(x)$, that

$$\|M_r^{\sharp,\mathcal{P}} f\|_{L^p} \le c \|M_{r,\nu}^{\sharp,\mathcal{P}} f\|_{L^p(\nu_p)},$$

for $p > 1$ when ν is doubling, or for $p > 0$ when $w \in A_\infty(\mu)$.
It also follows that

$$\|M_r^{\sharp,\mathcal{P}} f\|_{L^p(\nu_{1-p})} \le c \|M_{r,\nu}^{\sharp,\mathcal{P}} f\|_{L^p(\nu)}$$

provided that $w^{1-p} \in A_p(\mu)$, or equivalently, if $w \in A_{p'}(\mu)$, $1/p + 1/p' = 1$.
Finally, by Theorem 3 in Chapter I and Theorem 3, we conclude that if $w^{1-p} \in A_{p/r}(\mu)$, $p > r \ge 0$, then

$$\|M_{r,\nu}^{\sharp,\mathcal{P}} f\|_{L^p(\nu)} \le c \|f\|_{L^p(\nu_{1-p})}, \qquad (19)$$

and if $w \in A_{p'}(\mu)$, then there is a function $p_f \in \mathcal{P}$ such that

$$\|f - p_f\|_{L^p(\nu_{1-p})} \le c \|M_{r,\nu}^{\sharp,\mathcal{P}} f\|_{L^p(\nu)}. \qquad (20)$$

Inequalities (19) and (20) are important in later chapters when we consider the spaces $L_{p,\nu}^{\sharp,\mathcal{P}}$ consisting of those functions f, modulo $\mathcal{P}$, such that $M_{1,\nu}^{\sharp,\mathcal{P}} f \in L^p(\nu)$; (19) and (20) state that when $w \in A_{p'}(\mu)$, then $L_{p,\nu}^{\sharp,\mathcal{P}}$ coincides with $L^p(\nu_{1-p})$ modulo $\mathcal{P}$.

We consider now the control of the sharp maximal function by the local sharp maximal functions; this result corresponds to Proposition 4.

Proposition 6. Suppose ν is a weighted measure with respect to μ with w in $RH_{r_0}(\mu) \cap A_\infty(\mu)$, and $r_1 > r(r + r_0 - 1)/r_0$. Then,

$$M_{r,\nu}^{\sharp,\mathcal{P}} f(x) \le c M_\nu((M_{0,s}^{\sharp,\mathcal{P}} f)^{r_1})(x)^{1/r_1}.$$

Proof. Let $\mathcal{B} = \{B_j : B_j \subseteq B^1 \text{ and } A_0^P(f, B_j) > \alpha\}$. Note that the proof of Theorem 2 also gives

$$\mu\left(\{y \in B : |f(y) - p_B(y)| > \lambda\} \setminus \bigcup_{B_j \in \mathcal{B}} B_j\right) \le ce^{-c\lambda/\alpha}\mu(B), \quad \text{all } \lambda, \alpha > 0.$$

Now fix a ball B; we want to estimate

$$A_1 = \mu\left(\bigcup_{B_j \in \mathcal{B}} B_j\right)$$

in terms of

$$A^r = \frac{1}{\nu(B^1)} \int_{B^1} \left(M_{0,\nu,s}^{\sharp,P} f\right)^r d\nu.$$

By a covering argument we may assume that the B_j's are pairwise disjoint. Since we also have that $M_{0,\nu,s}^{\sharp,P} f(x) > \alpha\mu(B_j)/\nu(B_j)$ for all x in B_j, we conclude that with α_1 a number to be chosen shortly,

$$A_1 \le \sum_j \mu(B_j)$$

$$\le c\sum \mu(\{B_j : \nu(B_j)/\mu(B_j) > \alpha_1\}) + \alpha_1^{r_1-1}\sum (\mu(B_j)/\nu(B_j))^{r_1}\nu(B_j)$$

$$\le c\sum \mu(\{B_j : \nu(B_j)/\mu(B_j) > \alpha_1\}) + \alpha_1^{r_1-1}(A/\alpha)^{r_1}\nu(B^1).$$

Since $w \in RH_{r_0}(\mu)$ it follows that

$$\mu(\{x \in B : w(x) > \alpha_1\}) > \mu(B)/2, \quad \text{whenever} \quad \nu(B)/\mu(B) > \alpha_1.$$

Thus, the first sum above is dominated by

$$c\mu(\{x \in B^1 : w(x) > c\alpha_1\}) \le c\alpha_1^{-r_0}\int_{B^1} w(x)^{r_0} d\mu$$

$$\le c\alpha_1^{-r_0}\mu(B_1)(\nu(B^1)/\mu(B^1))^{r_0}.$$

It then follows that

$$A_1/\mu(B) \le c\alpha_1^{-r_0}(\nu(B^1)/\mu(B^1))^{r_0} + \alpha_1^{r_1-1}(A/\alpha)^{r_1}\nu(B)/\mu(B).$$

Whence by choosing

$$\alpha_1 = (\alpha/A)^{r_1/(r_0+r-1)}(\nu(B)/\mu(B))^{-(r_0-1)/(r_0+r-1)}$$

we get

$$A_1/\mu(B) \le c(A\nu(B)/\nu(B)\alpha)^{r_1 r_0/(r+r_0-1)}.$$

For $\varepsilon > 0$ a small number, we set $\alpha = \lambda^{1-\varepsilon} A\nu(B)/\mu(B)^\varepsilon$ and we note that

$$\mu(\{y \in B : |f(y) - p_B(y)| > \lambda\}) \le c(A\nu(B)/\lambda\mu(B))^{r_2}\mu(B),$$

provided that $r_2 < r_1 r/(r+1)$. Thus, if in addition $r < r_2$, it follows that

$$A_r^{\mathcal{P}}(f, B)^r \le \left(1 + \int_1^\infty \lambda^{r-r_2-1}d\lambda\right) \le c(A\nu(B)/\mu(B))^r.$$

Whence $(\mu(B)/\nu(B))A_r^{\mathcal{P}}(f, B) \le cA$, and we are done. ∎

Next we consider the dependence of the sharp maximal functions on $\mathcal{P}$; this result corresponds to Lemma 5. In the proof of that lemma there is a constant $\gamma > \gamma_0(\mu)$, where if λ denotes the Lebesgue measure we have $\gamma_0(\lambda) = 0$. The proof of the general result follows along similar lines, except that now we must handle terms involving $\mu(B)/\nu(B)$ in our estimates. Thus setting $B(x, 2^j) = B_j$, (16) in that lemma must be replaced by the following estimate: If $\Phi_k = (\nu(B_0)/\mu(B_0))((\mu(B_k)/\nu(B_k))A_r^{\mathcal{P}}$ and $\Psi_k = (\mu(B_{k-1})/\nu(B_{k-1}))A_r^{\mathcal{P}}(f, B_{k-1})$, then we have

$$\sup_{B(x,2^{\varepsilon k})} |p_{B_k}(f) - p_{B_{k-1}}(f) - p_k^0|$$

$$\le \left(\mu(B(x,2^{\varepsilon k}))/\mu(B_{k-1})\right)^\gamma \left(A_r^{\mathcal{P}}(f, B_k) + A_r^{\mathcal{P}}(f, B_{k-1})\right)$$

$$\le c \left(\mu(B(x,2^{\varepsilon k}))/\mu(B_0)\right)^\gamma \left(\mu(B_0)/\mu(B_k)\right)^{\gamma-q_0+1} \Phi_k + \Psi_k$$

$$\le c 2^{(c_1\varepsilon\gamma-c_2(\gamma-q_0+1))k}\Phi_k + \Psi_k.$$

In the above estimate we used the fact that $\mu \in D_{q_0}$. If $\gamma - q_0 + 1$ is positive and ε is small enough, the exponent of 2 in the above estimate is negative, and consequently, as in Lemma 5, the infinite sum converges uniformly on compact sets provided $M_{r,\nu}^{\sharp,\mathcal{P}} f$ is not identically ∞. In that proof we also replace (17) with the estimate

$$A_r^{\mathcal{P}_0}((f - p_f), B)\mu(B)/\nu(B) \le c \sup_{B_1 \supseteq B} A_r^{\mathcal{P}}(f, B_1)\mu(B_1)/\nu(B_1).$$

To obtain the above estimate we observe that the proof of (17) still works if we again invoke the fact that $\mu \in D_{q_0}$, provided $\gamma > q_0 - 1$. It then follows that if $\gamma = \gamma_0(\mu) = q_0 - 1$, $\mu \in D_{q_0}$, we have

$$M_{r,\nu}^{\sharp,\mathcal{P}_0}(f - p_f)(x) \le cM_{r,\nu}^{\sharp,\mathcal{P}} f(x), \quad \text{with } p_f \in \mathcal{P}. \tag{21}$$

The reader may profit from considering a particular case of the results discussed above, to wit, $X = R^n$, μ the Lebesgue measure, and $\mathcal{P}$ the class of polynomials of degree not exceeding m. In this context we have

Lemma 7. Suppose ν is a doubling weighted measure with respect to the Lebesgue measure on R^n with weight w, and let $p \in \mathcal{P}$. Then there is a constant $s_0 = s_0(n, m, w)$ with the following property: If $0 < s \leq s_0$, and the cube Q in R^n and number $A > 0$ are such that

$$\nu(\{x \in Q : |p(x)| > A\}) \leq s\nu(Q),$$

then we have $\sup_{x \in Q} |p(x)| \leq cA$.

Proof. Assume that p is not constant, pick $x_0 \in Q$ such that $|p(x_0)| = \sup_{x \in Q} |p(x)|$, and let $Q_0 \subset Q$ be a cube with sidelength half that of Q with x_0 as one of its corners. Since ν is doubling we have

$$\nu(\{x \in Q_0 : |p(x)| > A\}) \leq \nu(\{x \in Q : |p(x)| > A\}) \leq s\nu(Q) \leq sc_1\nu(Q_0).$$

By a change of coordinates and scaling if necessary we may assume that $x_0 = 0$ and $Q_0 = \{x \in R^n : 0 \leq x_i \leq 1, i = 1, \ldots, n\}$.

First we determine how far from $x_0 = 0$ are the zeros of p. Let x' denote a unit vector in R^n, tx' the vector emanating from the origin with length t in the direction of x', and set $I_0 = \{t \in R : tx' \in Q_0\}$. Then $p(t) = p(tx')$ is a polynomial of degree N, $0 \leq N \leq m$, in the variable t, and consequently, it can be factored $p(t) = a(t - a_1) \cdots (t - a_N)$; we have $\sup_{t \in I_0} |p(t)| = |aa_1 \cdots a_N|$. Observe that the length of the shortest among the vectors tx' with $t \in I_0$ is at least the sidelength of Q_0, i.e., 1. Moreover, even if all the zeros of $p(t)$ are equally distributed on such vector, which is the most unfavourable case in what follows, there is at least a value t_0 such that

$$|t_0 - a_j| > 1/N, \quad \text{all } N.$$

Clearly, $0 \leq t_0 \leq \sqrt{n}$. On the other hand, also note that

$$\frac{|p(t_0 x')|}{|p(0)|} = \frac{|t_0 - a_1| \cdots |t_0 - a_N|}{|a_1| \cdots |a_N|} \leq 1. \tag{22}$$

We are now interested in a lower bound for the left-hand side of (22). For each $0 \leq j \leq N$ we consider two cases, namely, $|a_j| \geq 2\sqrt{n}$, and $|a_j| < 2\sqrt{n}$. In the former case we have $|a_j| \leq |a_j - t_0| + |t_0| \leq |a_j - t_0| + \sqrt{n} \leq |a_j - t_0| + |a_j|/2$, and consequently,

$$|t_0 - a_j|/|a_j| \geq 1/2.$$

As for the latter case we have

$$|t_0 - a_j|/|a_j| \geq 1/(N2\sqrt{n}). \tag{23}$$

So, in either case (23) holds. Next we claim that

$$|t_0 - a_j|/|a_j| \geq (2\sqrt{n}N)^{N-1}, \quad \text{all } j. \tag{24}$$

For, if (24) does not hold for some j, (23) implies that (22) is contradicted.

Now, since also $|t_0 - a_j| > 1/N$, we finally obtain the estimates

$$|a_j| \geq 1/(2\sqrt{n}N)^{N-1} \quad \text{and} \quad |t_0 - a_j| > 1/N^N(2\sqrt{n})^N.$$

Suppose we restrict our attention to the range $0 \leq t \leq 1/2N^N(2\sqrt{n})^N$. Then as above we see that

$$|a_j| \leq |a_j - t| + 1/2N^N(2\sqrt{n})^N \leq |a_j - t| + |a_j|/2 \quad \text{or} \quad |a_j - t|/|a_j| > 1/2,$$

for all j. Whence it follows that

$$\frac{|p(tx')|}{|p(0)|} > 1/2^N \quad \text{whenever} \quad t \leq 1/2N^N(2\sqrt{n})^N.$$

Let $c^{-1} = 2m^m(2\sqrt{n})^m$. Since $N \leq m$ the above estimate reads

$$|p(tx')| > 2^{-m} \sup_{y \in Q_0} |p(y)| \quad \text{whenever} \quad t \leq c.$$

By varying x' over $|x'| = 1$, it follows that

$$|p(x)| > 2^{-m} \sup_{y \in Q_0} |p(y)| \quad \text{for} \quad x \in Q_0 \cap \{|x| \leq c\} = Q_0'.$$

Also, since ν is doubling, it follows that $\nu(Q_0) \leq c_2\nu(Q_0')$.

If A is so that the hypothesis hold, then

$$\nu(\{x \in Q_0 : |p(x)| > A\}) \leq sc_1\nu(Q_0) \leq sc_1c_2\nu(Q_0)$$
$$\leq sc_1c_2\nu(\{x \in Q_0 : |p(x)| > 2^{-m} \sup_{y \in Q_0} |p(y)|\}).$$

Suppose now that $A \leq 2^{-m} \sup_{y \in Q_0} |p(y)|$ and that $sc_1c_2 < 1$. Then the above estimate cannot hold and it follows that, under our assumptions, we must have

$$\sup_{y \in Q_0} |p(y)| \leq 2^m A. \quad \blacksquare$$

Corollary 8. When restricted to $\mathcal{P}$, all $L^p(\nu)$ norms on each cube Q are equivalent, $1 \leq p \leq \infty$.

Proof. For s_0 as in Lemma 7 and $p \in \mathcal{P}$, let

$$A = \frac{2}{s_0\nu(Q)} \int_Q |p(y)|w(y)\,dy.$$

Then we have

$$\nu(\{y \in Q : |p(y)| > A\}) \leq \int_Q |p(y)|w(y)\,dy \leq s_0\nu(Q)/2,$$

and from Lemma 7 it follows that $\sup_{y \in Q} |p(y)| \leq cA$, which gives the desired conclusion. $\blacksquare$

Lemma 9. Let $Q_0 \subseteq Q_1$ be cubes in R^n and suppose $p \in \mathcal{P}$. Then there is a constant $c = c(n, m)$ such that for any $x \in Q_0$ we have

$$\sup_{y \in Q_0} |p(y) - p(x)| \le c(|Q_0|/|Q_1|)^{1/n} \sup_{y \in Q_1} |p(y)|.$$

Proof. As we may assume that x is a corner of Q_0, we suppose that it is the farthermost from the center of Q_1. Let $R =$ sidelength of Q_1, and $r =$ sidelength of Q_0. Then the portion of the segment $x + t(y - x)$, $t > 0$, emanating from x in the direction of y totally contained in Q_1, corresponds to those values of t such that

1. $t \le 1$, because $y \in Q_0$, and

2. $|y + t(x - y)| \le R/2$, which implies that $t \le (R/|x - y|) + (2r/|x - y|)$, because this is the least favourable case.

At any rate, for values of $t \le c((R/r) + 1) \le (1/2)(|Q_1|/|Q_0|)^{1/n} = a$, say, the segment described above is totally contained in Q_1. Set now $p(t) = p(x + t(y - x))$. Then $p(1) = p(y)$, and

$$\sup_{0 < t < a} |p(t)| \le \sup_{y \in Q_1} |p(y)|.$$

Since we are now in a one-dimensional segment we reduce the proof to the case $n = 1$, $Q_0 = [0,1]$, $Q_1 = [0,a]$, and $p(0) = 0$. We can then write $p(t) = t p_1(t)$, $p_1(0) \ne 0$, and from Lemma 7 it follows that there is a number $t_0 > a(1 - s_0)$ such that $|p_1(t_0)| > \sup_{0 < t < a} |p_1(t)|/2c$, where s_0, c are the constants in that lemma. For, suppose this is not the case, and observe that when $t > a(1 - s_0)$, we have $|p_1(t)| \le \sup_{0 < t < a} |p_1(t)|/2c$, and consequently,

$$|\{t \in Q_1 : |p_1(t)| > \sup_{Q_1} |p_1|/2c\}| \le s_0 |Q_1|.$$

By Lemma 7 it then follows that

$$\sup_{t \in Q_1} |p_1(t)| \le c \sup_{t \in Q_1} |p_1(t)/2c| = \sup_{t \in Q_1} |p_1(t)|/2,$$

which is a contradiction. Consequently, for $0 < t_1 < 1$ we get

$$|p(t_1)| = t_1 |p_1(t_1)| \le t_1 2c |p_1(t_0)| \le (t_1/t_0) 2c |p(t_0)|$$
$$\le \frac{2c}{a(1 - s_0)} \sup_{t \in Q_1} |p(t)|.$$

Collecting these estimates and carrying them out for every segment in Q_0 we get

$$\sup_{y \in Q_0} |p(y) - p(x)| \le \frac{2c}{1 - s_0} 2(|Q_0|/|Q_1|)^{1/n} \sup_{y \in Q_1} |p(y)|,$$

and we are done. ∎

Lemmas 7 and 9 serve as a motivation for the assumptions in the general case. In the case of R^n, the results imply that the definition of $M_{r,\nu}^{\sharp,\mathcal{P}} f$ is independent of $\mathcal{P}$ provided that $m \geq m_0 = m_0(w)$. Later on, when we discuss the atomic decomposition of the weighted Hardy spaces, we will see that m_0 turns out to be directly related to the number of moments we need to impose on the atoms. Also the sharp maximal functions are important in establishing the continuity of singular integral and multiplier operators on these spaces. We finally remark that the condition $\gamma > q_0 - 1$ was used to get a pointwise estimate between sharp maximal functions. If we were merely interested in the equivalence of the $L^p(\nu)$ norms, then we could have chosen a possibly smaller γ depending, of course, on p and other conditions on w. In fact, if we let $d\nu_a(x) = w(x)^a d\mu$, $0 < a < 1$, we may replace $M_{r,\nu}^{\sharp,\mathcal{P}}$ by $M_{r,\nu_a}^{\sharp,\mathcal{P}}$ and w by w^{1-a} in estimates (15) and (16), and also use estimate (21) with ν replaced by ν_a.

Sources and Remarks.   The systematic use of the concept of median value for the study of properties related to the oscillation of a function goes back to F. John [1964] and more recently to J.-O. Strömberg [1979a], where the function $M_{0,\mu,s}^{\sharp}$ is introduced. The observation that for cubes in R^n and $\mathcal{P}$ the class of polynomials of degree less that or equal to m, the norms of $M_1^{\sharp,\mathcal{P}}$ and $M_1^{\sharp}$ are equivalent, is due to Berman [1975]. Theorem 2 is essentially the so-called John-Nirenberg inequality, cf. F. John and L. Nirenberg [1961], and Theorem 3 is an extension of a result due to C. Fefferman and E. Stein [1972]. Further properties of the local sharp maximal functions were explored by B. Jawerth and A. Torchinsky [1985].

CHAPTER

IV

Functions in the Upper

Half-Space

Let $F(y,t)$ be a continuous function defined in the upper half-space $R_{n+1}^{+} = \{(y,t) : y \in R^n, t > 0\}$. It is often of interest to study expressions involving F restricted to cones $\Gamma_a(x) = \{(y,t) : |x - y| < at\}$ with vertex at x and "opening" a. With this general situation in mind, and if χ denotes the characteristic function of the interval $[0,1]$, we introduce the functions $\varrho_{\lambda,a}(x,t)$ by

$$\varrho_{\lambda,a}(x,t) = \min\{1, (at/|x|)^{\lambda}\}, \quad a > 0, 1 \leq \lambda < \infty,$$

$$\varrho_{\infty,a}(x,t) = \chi(|x|/at), \quad a > 0,$$

and

$$\varrho_{\infty,0}(x,t) = \begin{cases} 0 & \text{if } x \neq 0 \\ 1 & \text{if } x = 0. \end{cases}$$

We consider, then, for the values of a and λ indicated above and for $0 < p < \infty$,

$$N_{\lambda,a,p}(F,x) = \left(\frac{1}{v} \int \int_{R_+^{n+1}} (\varrho_{\lambda,a}(x - y,t)F(y,t))^p (at)^{-n} dy \frac{dt}{t}\right)^{1/p},$$

where v denotes the volume of the unit ball in R^n, and

$$N_{\lambda,a,\infty}(F,x) = \sup_{(y,t) \in \Gamma_a(x)} |F(y,t)|, \quad p = \infty.$$

Our basic assumption is that $N_{\lambda,a,p}(F,x)$ is a Lebesgue measurable function of x and that it is lower semicontinuous when $p \neq \infty$. For a weighted measure ν with respect to the Lebesgue measure of R^n with weight w, we are interested in comparing the $L^q(\nu)$ norms of these functions for different values of a and λ. We always assume that ν is doubling, and to conform to tradition we use interchangeably the notations $L^q(\nu)$ and $L_w^q(R^n)$, or even L_w^q, and denote by $w \in A_p$, $1 \leq p \leq \infty$, the fact that w satisfies the A_p condition with respect to the Lebesgue measure.

If also $w \in A_\infty$ and F is the extension of a tempered distribution to the upper half-space by means of the convolution with the dilations of the heat kernel, it is then possible to obtain an equivalence of the $L_w^q(R^n)$ norms of the functions $N_{\infty,b,\infty}(F,x)$ and $N_{\infty,a,2}(t\frac{\partial}{\partial t}F,x)$. This is one of the basic results in the theory and it is discussed in Chapter VI.

We start with the case $p \neq \infty$, and discuss the case $p = \infty$ after the proof of Theorem 6. Let $N_{\infty,a,p}(F,x) = S_{a,p}(F,x)$ denote the Lusin, or area function, and $N_{\lambda,1,p}(F,x) = g^*_{\lambda,p}(F,x)$ denote the Littlewood-Paley function associated to F.

Since the techniques of proof differ for $0 < q \leq p$ and $p < q$, we state the corresponding results separately.

Theorem 1.   Suppose $\nu \in D_b$, and let $a \geq 1$ and $0 < q \leq p < \infty$. Then

$$\|S_{a,p}(F)\|_{L_w^q} \leq ca^{b/q - n/p}\|S_{1,p}(F)\|_{L_w^q}.$$

Theorem 2.   Suppose that $\nu \in D_b$, and let $a \geq 1$ and $q > p$. Then

$$\|S_{a,p}(F)\|_{L_w^q} \leq ca^{(b-n)/p}\|S_{1,p}(F)\|_{L_w^q}.$$

A sharpened version of this theorem is useful in applications. It is obtained by assuming that in addition $w \in A_r$.

Theorem 3.   Suppose that $\nu \in D_b$ and that $w \in A_r$, $1 \leq r < \infty$. If $a \geq 1$ and $0 < p < q$, we have
$$\|S_{a,p}(F)\|_{L_w^q} \leq ca^\gamma\|S_{1,p}(F)\|_{L_w^q},$$
where

$$\gamma = \begin{cases} 0 & \text{if } r \leq q/p \\ \left(1 - \frac{(q/p-1)}{r-1}\right)(b-n)/q & \text{if } q/p < r. \end{cases}$$

From these results we get

Theorem 4. Suppose $\nu \in D_b$. Then

$$\|g^*_{\lambda,p}(F)\|_{L^q_w} \leq c\|S_{1,p}(F)\|_{L^q_w},$$

provided that either $q \leq p$ and $\lambda > b/q$, or $p < q$ and $\lambda > b/p$.

Theorem 5. Suppose that $\nu \in D_b$ and $w \in A_r$, $1 \leq r < \infty$. If $q > p$, the conclusion of Theorem 4 holds provided that

$$\lambda > \begin{cases} n/p & \text{if } q/p \geq r \\ n/p + \left(1 - \frac{(q/p-1)}{r-1}\right)(b-n)p/q & \text{if } q/p < r. \end{cases}$$

Theorem 6. Suppose $\nu \in D_b$, and that $0 < q < p$ and $\lambda = b/q$. Then the following weak-type inequality holds:

$$\nu(\{g^*_{\lambda,p}(F) > s\}) \leq cs^{-q}\|S_{1,p}(F)\|^q_{L^q_w}, \quad \text{all } s > 0.$$

We begin by describing several facts, essentially of geometric nature, describing the saw-tooth regions in the upper half-space which appear as we consider the integral of $S_{a,p}(F)^q$ over subsets of R^n.

Lemma 7. Suppose $\nu \in D_b$ is a doubling measure with constant c_d, i.e., $\nu(B(x,rt)) \leq c_d t^b \nu(B(x,r))$ for all $x \in R^n$, $r > 0$ and $t \geq 1$. Given an open set $\mathcal{O}$ in R^n, we associate with it the set $\mathcal{U}$ defined by

$$\mathcal{U} = \{x \in R^n : M_\nu \chi_\mathcal{O}(x) > (2c_d a^b)^{-1}\}, \quad 0 < a < \infty.$$

Then
   (i) $\Gamma_a(\mathcal{U}^c) = \bigcup_{x \in \mathcal{U}^c} \Gamma_a(x) \subseteq \Gamma_1(\mathcal{O}^c) = \bigcup_{x \in \mathcal{O}^c} \Gamma_1(x)$.
   (ii) If $(y,t) \in \Gamma_a(\mathcal{U}^c)$, then $\nu(B(y,t)) \leq 2\nu(B(y,t) \cap \mathcal{O}^c)$.

Proof. If $(y,t) \in \Gamma_a(\mathcal{U}^c)$, there is $x \in \mathcal{U}^c$ with $|x - y| < at$, and therefore $x \in B(y,at)$. Thus,

$$\nu(B(y,t) \cap \mathcal{O})/\nu(B(y,t)) \leq c_d a^b \nu(B(y,at) \cap \mathcal{O})\nu(B(y,at))$$
$$\leq c_d a^b M_\nu \chi_\mathcal{O}(x) \leq c_d a^b (2c_d a^b)^{-1} \leq 1/2.$$

From this estimate (ii) follows at once. Furthermore, if $(y,t) \in \Gamma_a(\mathcal{U}^c)$, by (ii) we conclude that $B(y,t) \cap \mathcal{O}^c \neq \emptyset$, and if $x \in B(y,t) \cap \mathcal{O}^c$ it follows that $(y,t) \in \Gamma_1(x)$, and so $(y,t) \in \Gamma_1(\mathcal{O}^c)$, and (i) holds as well. ∎

Lemm 8. Let $\nu \in D_b$ with weight w, and $a > 1$. If $\mathcal{O}$ is an open set of R^n of finite measure and if $\mathcal{U}$ is associated to $\mathcal{O}$ as in Lemma 7, we have

$$\int_{\mathcal{U}^c} S_{a,p}(F,x)^p w(x)\,dx \le 2c_d a^{b-n} \int_{\mathcal{O}^c} S_{1,p}(F,x)^p(x)w(x)\,dx.$$

Proof. From the definition of the Lusin function it follows that

$$\int_{\mathcal{U}^c} S_{a,p}(F,x)^p w(x)\,dx = \int\int_{\Gamma_a(\mathcal{U}^c)} |F(y,t)|^p \nu(B(y,at)\cap \mathcal{U}^c)\frac{1}{v(at)^n}\,dy\frac{dt}{t},$$

and

$$\int_{\mathcal{O}^c} S_{1,p}(F,x)^p w(x)\,dx = \int\int_{\Gamma_1(\mathcal{O}^c)} |F(y,t)|^p \nu(B(y,t)\cap \mathcal{O}^c)\frac{1}{vt^n}\,dy\frac{dt}{t}.$$

Since, by Lemma 7, $\Gamma_a(\mathcal{U}^c) \subseteq \Gamma_1(\mathcal{O}^c)$ and

$$\nu(B(y,at)\cap\mathcal{U}^c) \le \nu(B(y,at)) \le c_d a^b \nu(B(y,t)) \le 2c_d a^b \nu(B(y,t)\cap\mathcal{O}^c),$$

the conclusion follows. ∎

We prove now a result of different nature, and once this is done we will be ready to prove Theorem 1.

Lemma 9. Let ν be a weighted measure with respect to the Lebesgue measure on R^n with weight w, and f,g be nonnegative functions on R^n. For $s > 0$ let $E_s = \{f > s\}$ and $E'_s = \{g > s\}$. If $g \in L^p_w(R^n)$, $0 < q < p$, and

$$\nu(E_{rs}) \le c_1\frac{1}{(rs)^p}\int_0^s t^{p-1}\nu(E'_t)\,dt + c_2\nu(E'_s), \quad r,s > 0$$

then $f \in L^q_w(R^n)$, and

$$\|f\|_{L^q_w} \le c\|g\|_{L^q_w}, \quad c = 2c_1^{1/p}c_2^{1/q-q/p}(p-q)^{-1/p}.$$

Proof. Since $\|f\|_{L^q_w}^q = q\int_0^\infty (rs)^{q-1}\nu(E_{rs})\,d(rs)$, we have

$$\|f\|_{L^q_w}^q \le q\int_0^\infty (rs)^{q-1}\left((c_1/(rs)^p)\int_0^s t^{p-1}\nu(E'_t)\,dt\right)d(rs)$$

$$+ qc_2\int_0^\infty (rs)^{q-1}\nu(E'_s)\,d(rs) = I + J,$$

say. Moreover, since $J = c_2 r^q \|g\|_{L_w^q}^q$, and

$$I \leq c_1 q \int_0^\infty t^{p-1} \nu(E_t') \int_t^\infty (rs)^{q-p-1} d(rs) dt$$

$$= c_1 q r^{q-p} \frac{1}{p-q} \int_0^\infty t^{q-1} \nu(E_t') dt = c_1 \frac{r^{q-p}}{p-q} \|g\|_{L_w^q}^q,$$

the conclusion follows upon setting $r = (c_1/c_2)^{1/p}(p-q)^{-1/p}$. ∎

Proof of Theorem 1. Let $\mathcal{O}_s$ be the open set of finite measure $\mathcal{O}_s = \{S_{1,p}(F) > s\}$, and let $\mathcal{U}_s$ be associated to $\mathcal{O}_s$ as in Lemma 7. Further let $\mathcal{O}_s' = \{S_{a,p}(F) > s\}$.

Now, for all $r, s > 0$ we have $\nu(\mathcal{O}_{sr}') \leq \nu(\mathcal{U}_s) + \nu(\mathcal{O}_{sr}' \cap \mathcal{U}_s^c)$. Note that from Lemma 8 it follows that

$$\nu(\mathcal{O}_{rs}' \cap \mathcal{U}_s^c) \leq (sr)^{-p} \int_{\mathcal{U}_s^c} S_{a,p}(F, x)^p w(x) \, dx$$

$$\leq c a^{b-n} (sr)^{-p} \int_{\mathcal{O}_s^c} S_{1,p}(F, x)^p w(x) \, dx$$

$$\leq c a^{b-n} (sr)^{-p} p \int_0^s t^{p-1} \nu(\mathcal{O}_t) \, dt.$$

Also, since M_ν is of weak-type $(1,1)$ on $L^1(\nu)$, it follows that

$$\nu(\mathcal{U}_s) \leq c a^b \nu(\mathcal{O}_s).$$

We combine these estimates and invoke Lemma 9 with $c_1 = c a^{b-n}$ and $c_2 = c a^b$ to get

$$\|S_{a,p}(F)\|_{L_w^q}^q \leq c a^{(b-n)q/p} a^{b(1-q/p)} (p-q)^{-q/p} \|S_{1,p}(F)\|_{L_w^q}^q$$

$$= c a^{b-nq/p} (p-q)^{-q/p} \|S_{1,p}(F)\|_{L_w^q}^q.$$

Finally, when $p = q$ we have

$$\|S_{a,p}(F)\|_{L_w^p}^p = \int \int_{R_{n+1}^+} |F(y,t)|^p \nu(B(y,at)) \frac{1}{v(at)^n} dy \frac{dt}{t}$$

$$\leq c_d a^{b-n} \int \int_{R_{n+1}^+} |F(y,t)|^p \nu(B(y,t)) \frac{1}{vt^n} dt \frac{dt}{t}$$

$$= c_d a^{b-n} \|S_{1,p}(F)\|_{L_w^p}^p. \quad ∎$$

Proof of Theorem 2. Since $q > p$, to estimate $\|S_{a,p}(F)\|_{L_w^q}^q$ we use the converse to Hölder's inequality and compare the integrals

$$I = \int_{R^n} S_{a,p}(F,x)^p g(x) w(x)\, dx, \quad \text{and} \quad J = \int_{R^n} S_{1,p}(F,x)^p M_\nu g(x) w(x)\, dx,$$

where $g \in L_w^{(q/p)'}(R^n)$ has norm less than or equal to 1. Since

$$M_g(z) \geq \frac{1}{\nu(B(y,at))} \int_{B(y,at)} g(x) w(x)\, dx, \quad \text{all} \quad z \in B(y,t)$$

we have

$$\int_{B(y,at)} g(x) w(x)\, dx \leq ca^b \frac{\nu(B(y,t))}{\nu(B(y,at))} \int_{B(y,at)} g(x) w(x)\, dx$$

$$\leq ca^b \int_{B(y,t)} M_\nu g(x) w(x)\, dx,$$

and consequently,

$$I \leq ca^{b-n} J \leq ca^{b-n} \|S_{1,p}(F)\|_{L_w^q}^p \|M_\nu g\|_{L_w^{(q/p)'}}.$$

Moreover, since ν is doubling, by Theorem 1 in Chapter I,

$$\|M_\nu g\|_{L_w^{(q/p)'}} \leq c\|g\|_{L_w^{(q/p)'}} \leq c.$$

Combining these estimates we get

$$\int_{R^n} S_{a,p}(F,x)^p g(x) w(x)\, dx \leq ca^{b-n} \|S_{1,p}(F)\|_{L_w^q}^p,$$

which gives the conclusion. ∎

Proof of Theorem 3. Since it follows along the lines to that of Theorem 2, we only sketch it. Suppose first that $q/p < r$ and set $d\nu_\varepsilon(x) = w(x)^\varepsilon dx$, where $0 < \varepsilon < p/q$ will be chosen shortly. Since

$$(\nu(B)/|B|)^\varepsilon \sim \nu_\varepsilon(B)/|B| \quad \text{all balls } B,$$

it follows that $\nu_\varepsilon \in D_{\varepsilon(b-n)+n}$ whenever $\nu \in D_b$.

Let $\delta = (\varepsilon - p/q)(q/p)'$ and $d\nu_\delta(x) = w(x)^\delta dx$. For $g \in L_{\nu_\delta}^{(q/p)'}(R^n)$ with norm less than equal to 1, we have

$$\int_{R^n} S_{a,p}(F,x)^p g(x) w(x)^\varepsilon dx \leq ca^{\varepsilon(b-n)} \int_{R^n} S_{1,p}(F,x)^p(x) M_{\nu_\varepsilon} g(x) w(x)^\varepsilon dx$$

$$\leq ca^{\varepsilon(b-n)} \|S_{1,p}(F)\|_{L_w^q}^p \|M_{\nu_\varepsilon} g\|_{L_{w^\delta}^{(q/p)'}}.$$

We choose now

$$\varepsilon = \left(1 - \frac{(q/p - 1)}{(r - 1)}\right)(p/q),$$

and observe that if $w \in A_r$, then $(w^\delta/w^\varepsilon) \in A_{(q/p)'}(\nu_\varepsilon)$. Whence, from the maximal theorem we get $\|M_{\nu_\varepsilon} g\|_{L_{w^\delta}^{(q/p)'}} \leq c\|g\|_{L_{w^\delta}^{(q/p)'}} \leq c$, and the conclusion follows in this case. When $q/p \geq r$, we choose $\varepsilon = 0$ and repeat the above argument. ∎

Proof of Theorem 4. We do the case $q \leq p$ first. Note that

$$\varrho_{1,\lambda}(x - y, t)^p \sim \sum_{k=0}^{\infty} 2^{-k\lambda p} \chi(|x - y|/2^k t).$$

Whence, multiplying this relation through by $|F(y,t)|^p (vt^n)^{-1}$ and integrating over R_{n+1}^+ with respect to $dy\frac{dt}{t}$, it follows that

$$g_{\lambda,p}^*(F, x)^p \leq c \sum_{k=0}^{\infty} 2^{-k(\lambda p - n)} S_{2^k,p}(F, x)^p.$$

Since $q/p \leq 1$, from Theorem 1 we get that

$$\|g_{\lambda,p}^*(F)\|_{L_w^q} \leq c \left(\int_{R^n} \left(\sum_{k \geq 0} 2^{-k(\lambda p - n)} S_{2^k,p}(F, x)^p\right)^{q/p} w(x)\,dx\right)^{1/q}$$

$$\leq c \left(\int_{R^n} \sum_{k \geq 0} 2^{-k(\lambda p - n)(q/p)} S_{2^k,p}(F, x)^q w(x)\,dx\right)^{1/q}$$

$$= c \left(\sum_{k \geq 0} 2^{-k(\lambda p - n)(q/p)} \|S_{2^k,p}(F)\|_{L_w^q}^q\right)^{1/q}$$

$$\leq c \left(\sum_{k \geq 0} 2^{-k(\lambda p - n)(q/p)} 2^{k(b - nq/p)}\right)^{1/q} \|S_{1,p}(F)\|_{L_w^q}.$$

This sum is finite provided that $\lambda q > p$, which is our assumption. The proof of the case $q > p$, being similar to that of Theorem 5 is left to the reader. ∎

Proof of Theorem 5. We only sketch it. We now have that $q/p > 1$, and consequently,

$$I = \|g_{\lambda,p}^*(F)\|_{L_w^q}^p = \|g_{\lambda,p}^*(F)^p\|_{L_w^{q/p}} \leq c \sum_{k \geq 0} 2^{-k(\lambda p - n)} \|S_2$$

By Theorem 3 it then follows that

$$I \leq c\Big(\sum_{k\geq 0} 2^{-k(\lambda p-n)}\Big)\|S_{1,p}(F)\|^p_{L^q_w}, \quad \text{if } q/p \geq r,$$

and, in case $q/p < r$, that

$$I \leq c\Big(\sum_{k\geq 0} 2^{-k(\lambda p-n)}2^{k(1-((q/p)-1)/(r-1))(b-n)(p/q)}\Big)\|S_{1,p}(F)\|^p_{L^q_w}.$$

In both instances our assumptions imply that the sums converge. ∎

Proof of Theorem 6. For $s > 0$ let $\mathcal{E} = \{M_\nu(S_{1,p}(F)^q) > s^q/2c_d\}$, and for $k \geq 0$ let

$$\mathcal{O}_k = \{S_{1,p}(F) > 2^{bk/q}s\}, \quad \text{and} \quad \mathcal{U}_k = \{M_\nu \chi_{\mathcal{O}_k} > (2c_d 2^{kb})^{-1}\}.$$

Observe that $\mathcal{U}_k \subseteq \mathcal{E}$ for all k. Indeed, if $x \in \mathcal{U}_k$, then there is a ball $B(y,r)$, say, which contains x and such that $\nu(B(y,r) \cap \mathcal{O}_k) > (2c_d 2^{kb})^{-1}\nu(B(y,r))$. Thus

$$\int_{B(y,r)} S_{1,p}(F,z)^q w(z)\,dz \geq \int_{B(y,r)\cap\mathcal{O}_k} S_{1,p}(F,z)^q w(z)\,dz$$

$$\geq 2^{kb}s^q\nu(B(y,r)\cap\mathcal{O}_k) \geq (2c_d)^{-1}s^q\nu(B(y,r)),$$

and $x \in \mathcal{E}$.

We also have

$$I = s^p\nu(\{g^*_{\lambda,p}(F) > s\} \cap \mathcal{E}^c) \leq \int_{\mathcal{E}^c} g^*_{\lambda,p}(F,x)^p w(x)\,dx$$

$$\leq c\sum_{k\geq 0} 2^{-k(\lambda p-n)}\int_{\mathcal{E}^c} S_{2^k,p}(F,x)^p w(x)\,dx$$

$$\leq c\sum_{k\geq 0} 2^{-k(\lambda p-n)}\int_{\mathcal{U}^c_k} S_{2^k,p}(F,x)^p w(x)\,dx$$

$$\leq c\sum_{k\geq 0} 2^{-k(\lambda p-n)+k(b-n)}\int_{\mathcal{O}^c_k} S_{1,p}(F,x)^p w(x)\,dx$$

$$\leq c\int_{R^n} S_{1,p}(F,x)^p\Big(\sum_{k\geq 0} 2^{-k(\lambda p-b)}\chi_{\mathcal{O}^c_k}(x)\Big) w(x)\,dx.$$

Let now h be the least nonnegative integer such that $x \in \mathcal{O}^c_k$. Then the above sum is of order $\sum_{k\geq h} 2^{-k(\lambda p-b)} \sim 2^{-h(\lambda p-b)}$. From the definition of h we have $S_{1,p}(F,x) \leq 2^{bh/q}s$, and consequently,

$$2^{-h(\lambda p-b)} \leq c(S_{1,p}(F,x)/s)^{-q(\lambda p-b)/b}.$$

Thus, since $p - q(\lambda p - b)/b = q$ when $\lambda = b/q$, it follows that

$$I \leq c \int_{R^n} S_{1,p}(F,x)^p \left(S_{1,p}(F,x)/s\right)^{-q(\lambda p - b)/b} w(x)\,dx$$

$$= cs^{q(\lambda p - b)/b} \int_{R^n} S_{1,p}(F,x)^q w(x)\,dx\,.$$

Since also

$$\nu(\{g^*_{\lambda,p}(F) > s\} \cap \mathcal{E}) \leq \nu(\mathcal{E}) \leq cs^{-q}\|S_{1,p}(F)\|^q_{L^q_w}\,,$$

the conclusion follows. ∎

The results for $p = \infty$ and $0 \leq a < \infty$ require different techniques of proof and are somewhat simpler. Let $N_{\infty,a,\infty}(F,x) = M_a(F,x)$ denote the non-tangential maximal function, $N_{\infty,0,\infty}(F,x) = M_0(F,x)$ denote the radial, and $N_{\lambda,1,\infty}(F,x) = N_\lambda(F,x)$ denote the tangential maximal function, respectively, associated to F. The following three results then hold.

Theorem 10. Let $\nu \in D_b$, $0 < a_1 < a_2 < \infty$, and suppose that $M_{a_1}(F)$ belongs to $L^q_w(R^n)$. Then also $M_{a_2}(F) \in L^q_w(R^n)$ and there is a constant c independent of F, a_1 and a_2 such that

$$\|M_{a_2}(F)\|_{L^q_w} \leq c(a_2/a_1)^{b/q}\|M_{a_1}(F)\|_{L^q_w}\,.$$

Theorem 11. Let $\nu \in D_b$, $0 < q < \infty$, and suppose $\lambda > b/q$. Then if $M_1(F)$ is in $L^q_w(R^n)$, also $N_\lambda(F) \in L^q_w(R^n)$ and there is a constant c independent of F such that

$$\|N_\lambda(F)\|_{L^q_w} \leq c\|M_1(F)\|_{L^q_w}\,.$$

Moreover, if $\lambda = b/q$ the following weak-type estimate holds:

$$\nu(\{N_\lambda > s\}) \leq cs^{-q}\|M_1(F)\|^q_{L^q_w}\,, \quad c = c(\lambda)\,.$$

Theorem 12. Let $\nu \in D_b$, and suppose $F(x,t)$ is continuously differentiable with respect to the x variables in $t > 0$. Let $G(x,t) = |\nabla F(x,t)|$, where

$$\nabla F(x,t) = \left\langle \frac{\partial}{\partial x_1}F(x,t),\ldots,\frac{\partial}{\partial x_n}F(x,t) \right\rangle,$$

is the gradient of F with respect to the x variables. Further, suppose that $M_0(F), M_{a_2}(G) \in L^p_w(R^n)$, $0 < a_1, a_2 < \infty$, and $0 < p < \infty$. Then there is a constant c depending only on a_1, a_2, b and p such that

$$\|M_{a_1}(F)\|_{L^p_w} \leq c\|M_0(F)\|^{p/(p+b)}_{L^p_w}\|M_{a_2}(G)\|^{b/(p+b)}_{L^p_w}\,,$$

if $\|M_0(F)\|_{L^p_w} \leq \|M_{a_2}(G)\|_{L^p_w}$, and

$$\|M_{a_1}(F)\|_{L^p_w} \leq c\|M_0(F)\|_{L^p_w}\,, \quad \text{otherwise}\,.$$

We begin by discussing a preliminary result.

Lemma 13. Let $\nu \in D_b$ and assume that $0 < a_1 < a_2 < \infty$. Then there is a constant c independent of F, a_1, a_2 such that for all $s > 0$,

$$\nu(\{M_{a_2}(F) > s\}) \leq c(a_2/a_1)^b \nu(\{M_{a_1}(F) > s\}).$$

Proof. Let $x \in \mathcal{O}_2 = \{M_{a_2}(F) > s\}$. Then there is (y,t) such that $|x - y| < a_2 t$ and $|F(y,t)| > s$; consequently, $B(y,at) \subseteq \{M_{a_1}(F) > s\} = \mathcal{O}_1$. We claim that

$$\mathcal{O}_2 \subseteq \mathcal{U} = \{M_\nu \chi_{\mathcal{O}_1}(x) > c_d^{-1}(a_1/(a_1 + a_2))^b\}.$$

Indeed, for $x \in \mathcal{O}_2$ we have

$$M_\nu \chi_{\mathcal{O}_1}(x) \geq \frac{1}{\nu(B(y,(a_1 + a_2)t))} \int_{B(y,(a_1+a_2)t)} \chi_{\mathcal{O}_1}(x) w(x) \, dx$$
$$\geq \nu(B(y,a_1 t))/\nu(B(y,(a_1 + a_2)t)) \geq c_d^{-1}(a_1/(a_1 + a_2))^b.$$

Whence, by the weak-type $(1,1)$ of M_ν on $L^1(\nu)$, it follows that

$$\nu(\mathcal{O}_2) \leq \nu(\mathcal{U}) \leq c(1 + (a_1/a_2))^b \nu(\mathcal{O}_1). \quad \blacksquare$$

Theorem 10 follows at once from Lemma 13. As for Theorem 11, we note that

$$N_\lambda(F,x) \leq c \sup_{k \geq 0} \left(2^{-\lambda k} M_{2^k}(F,x)\right).$$

Consequently, $\{N_\lambda(F) > s\} \subseteq \bigcup_{k \geq 0} \{M_{2^k}(F) > c2^{\lambda k}s\}$, and by Lemma 13,

$$\nu(\{N_\lambda(F) > s\}) \leq \sum_{k \geq 0} \nu(\{M_{2^k}(F) > c2^{\lambda k}s\})$$
$$\leq c \sum_{k \geq 0} 2^{kb} \nu(\{M_1(F) > c2^{\lambda k}s\}).$$

Suppose now that $\lambda = b/q$. We then have

$$2^{\lambda k q} s^q \nu(\{M_1(F) > c2^{\lambda k}s\}) \leq c \int_{c2^{(k-1)\lambda}s}^{c2^{k\lambda}s} s^{q-1} \nu(\{M_1(F) > s\}) \, ds.$$

Whence, substituing in the above sum, and since $\lambda q = b$, we get

$$\nu(\{N_\lambda(F) > s\}) \leq s^{-q} \sum_{k \geq 0} \int_{c2^{(k-1)\lambda}s}^{s2^{k\lambda}s} s^{q-1} \nu(\{M_1(F) > s\}) \, ds$$
$$\leq cs^{-q} \|M_1(F)\|_{L_w^q}^q.$$

If instead $\lambda > b/q$, we have

$$q \int_0^\infty s^{q-1} \nu(\{N_\lambda(F) > s\})\, ds \leq c\Big(\sum_{k \geq 0} 2^{k(b-\lambda q)}\Big) q \int_0^\infty \nu(\{M_1(F) > s\})\, ds$$
$$= c\|M_1(F)\|_{L_w^q}^q,$$

which is the desired conclusion.

Proof of Theorem 12. Without loss of generality we may assume that $a_1 = 1$ and $a_2 = 2$. Given $0 < r < 1$ and $s > 0$, consider the set

$$\mathcal{U} = \{M_1(F) > s\} \cap \{M_2(G) \leq r^{-1/p}s\}.$$

We claim that for a constant c independent of F, G, r, s, b and p we have

$$\nu(\mathcal{U}) \leq cr^{-b/p}\nu(\{M_0(F) > s/2\}).$$

If this is the case, we have

$$\nu(\{M_1(F) > s\}) \leq cr^{-b/p}\nu(\{M_0(F) > s/2\}) + \nu(\{M_2(G) > r^{-1/p}s\}).$$

Thus, multiplying through by ps^{p-1} and integrating we get

$$\|M_1(F)\|_{L_w^p}^p \leq cr^{-b/p}2^p\|M_0(F)\|_{L_w^p}^p + r\|M_2(G)\|_{L_w^p}^p.$$

If $I = \|M_0(F)\|_{L_w^p} \geq J = \|M_2(G)\|_{L_w^p}$, we set $r = 1$ above and replace J by I. If, on the other hand, $I \leq J$, we put $r = (I/J)^{p^2/(b+p)}$ and obtain the desired conclusion in this case as well.

So, it only remains to estimate $\nu(\mathcal{U})$. Let $M_1^k(F, x) = M_1(\chi_{[0,k]}F, x)$. Since $M_1^k(F, x)$ increases to $M_1(F, x)$ as $k \to \infty$, it suffices to work with $M_1^k(F)$ instead. Suppose, then, that $M_1^k(F, x) > s$. Then there is (y, t) such that $|x - y| \leq n$ and $|F(y, t)| > s$. If in addition $M_2(G, x) \leq r^{-1/p}s$, then we have $|G(z, t)| \leq r^{-1/p}s$ for $|x - z| \leq 2t$. From the mean value theorem it follows that $|F(y', t)| > s/2$ whenever $|y - y'| \leq r^{1/p}t/2$, and consequently, $M_0(F, y') > s/2$ for y' in $B(y, r^{1/p}t/2)$. Let $\{B(y_i, t_i)\}$ be a pairwise disjoint collection of these balls so that $\mathcal{U} \subseteq \bigcup_i B^1(y_i, t_i)$. Now, note that for any $x \in B(y_i, t_i)$ we have

$$B(x, 2t_i(1 + r^{1/p}/2)) \supseteq B(y_i, t_i(1 + r^{1/p}/2)).$$

Select $x_i \in B(y_i, t_i)$ such that $M_0(F, y') > s/2$ for y' in $B(x_i, r^{1/p}t_i/2)$. With this choice of x_i we get

$$\nu(\mathcal{U}) \leq c\sum \nu(B(y_i, t_i)) \leq c\sum \nu(B(x_i, 2t_i(1 + r^{1/p}/2)))$$
$$\leq c\sum \nu\Big(B\Big(x_i, \frac{2t_i(1 + r^{1/p}/2)t_i r^{1/p}/2}{t_i r^{1/p}/2}\Big)\Big)$$
$$\leq r^{-b/p}\sum \nu(B(x_i, t_i r^{1/p}/2)) \leq cr^{-b/p}\nu(\{M_0(F) > s/2\}).$$

Since all the constants above are independent of k, we are done. ∎

Sources and Remarks. The results in this chapter, as well as those in the rest of these notes, hold, with straightforward modifications in the proofs, in the so-called nonisotropic case. Specifically, let A_t, $t > 0$, $A_{ts} = A_t A_s$, be a continuous group of affine transformations of R^n leaving the origin fixed. If P denotes the infinitesimal generator of A_t, we have

$$t\frac{d}{dt}A_t = PA_t, \quad t > 0.$$

We further assume that if $|x|$ denotes the Euclidean norm, we have

$$t^\alpha|x| \leq |A_t x| \leq t^\beta|x|, \quad 1 \leq \alpha \leq \beta, \, t \geq 1,$$

and $\langle Px, x \rangle \geq \langle x, x \rangle$. Under these conditions, given $x \in R^n$, there exists a unique $t > 0$ such that $|A_t^{-1}x| = 1$. We then define $\rho(x)$ to be this value of t, and observe that ρ defines a translation invariant, nonisotropic if $P \neq I$, distance $\rho(x - y)$. Further details about this metric, and the general outline of this chapter, consult the paper by A. Calderón and A. Torchinsky [1975]. Theorem 6 is due to N. Aguilera and C. Segovia [1977]. As for the the tent spaces, cf. R. Coifman, Y. Meyer and E. Stein [1983]. In the case when $F(x, t) = f * P_t(x)$ is the convolution of a tempered distribution f with the Poisson kernel P_t of the upper half-space, some of these results are analogous to those of D. Burkholder and R. Gundy [1972], and C. Fefferman and E. Stein [1972]. The reader should consult these authors for further references concerning the unweighted case, i.e., $w(x) = 1$. For instance, in that setting Theorem 6 is due to C. Fefferman [1970], and was extended to the weighted case by B. Muckenhoupt and R. Wheeden [1974]. A weaker version of the result, in the context of our presentation, is due to A. Torchinsky [1979].

CHAPTER

V

Extensions of Distributions

In this chapter we consider tempered distributions $f \in \mathcal{S}'(R^n)$ on R^n and their extensions to the upper half-space R_+^{n+1}. Let ϕ be a function in the Schwartz class $\mathcal{S}(R^n)$, i.e., the Schwartz norms $|\phi|_{\alpha,\beta} = \sup_x |x^\alpha D^\beta \phi(x)|$ are finite for all multi-indices α, β of nonnegative integers; as usual, if $x = (x_1, \ldots, x_n)$ and $\alpha = (\alpha_1, \ldots, \alpha_n)$ and $\beta = (\beta_1, \ldots, \beta_n)$, then $|\beta| = \beta_1 + \cdots + \beta_n$, $x^\alpha = (x_1^{\alpha_1}, \ldots, x_n^{\alpha_n})$ and $D^\beta \phi(x) = \frac{\partial^{|\beta|}}{\partial x_1^{\beta_1} \ldots \partial x_n^{\beta_n}} \phi(x)$.

Further, let $\langle f, \phi \rangle$ denote the evaluation of the tempered distribution f at the test function ϕ, and $f * \phi_t(x)$ denote the $C^\infty(R^n)$ function $\langle f, \phi_t(x - \cdot) \rangle$; similar definitions may be given for the nonisotropic dilations discussed in Chapter IV. The extensions of f to R_+^{n+1} of interest to us are defined by

$$F_\phi(x,t) = f * \phi_t(x), \quad x \in R^n, \ t > 0.$$

We want to study how these extensions depend on ϕ. For instance, we will see that if $\phi \in \mathcal{S}(R^n)$ has a nonvanishing integral, then any extension F_ψ of f with $\psi \in \mathcal{S}$, is dominated at a point $(x,t) \in R_+^{n+1}$ by a constant $c_{\phi,\psi}$ independent of f times a weighted average of $|F_\phi|$ concentrated about the point (x,t). As an application of this result we get estimates between $N_{\lambda,a}(F_\phi)$'s for different ϕ's for some of the operators introduced in the preceding chapter.

Some estimates are restricted to the case when f is a function in the class $\hat{\mathcal{D}}_0$ defined as follows. Let $\mathcal{D} = \{\phi \in \mathcal{S}(R^n) : \text{supp } \phi \text{ is compact}\}$, and set $\hat{\mathcal{D}}_0 = \{\phi \in \mathcal{S}(R^n) : \hat{\phi} \in \mathcal{D} \text{ and } \hat{\phi} \text{ vanishes in a neighbourhood of the origin}\}$. As usual, for $\phi \in L^1(R^n)$, $\hat{\phi}(\xi) = \int_{R^n} e^{-ix\cdot\xi} \phi(x)\, dx$ denotes the Fourier transform of ϕ.

Also recall that for $g \in \mathcal{S}'(R^n)$ we have

$$\langle g_t, \phi \rangle = \langle g, \phi(t\cdot) \rangle, \quad \text{and} \quad \langle \hat{g}, \phi \rangle = \langle g, \hat{\phi} \rangle, \quad \text{all } \phi \in \mathcal{S}(R^n).$$

We are now ready to state the results.

Theorem 1. Let $\phi, \psi \in \mathcal{S}(R^n)$, and suppose that ϕ has nonvanishing integral. Then, for all p, $0 < p < \infty$, and every $a > 0$ (small) and $N > 0$ (large), there is a constant c which depends only on ϕ, ψ, p, a, N such that for all tempered distributions f and each $(x, t) \in R^{n+1}_+$ we have

$$|F_\psi(x,t)| \le C \int_0^{at} \int_{R^n} |F_\phi(y,s)|^p (1 + (|x-y|/s))^{-pN} s^{-n} (s/t)^{pN} dy \frac{ds}{s}.$$

In applications it is important to relax the assumptions on ϕ. A way to go about this is the following: Let $k \ge 1$, and suppose $\phi = (\phi^{(1)}, \dots, \phi^{(k)})$ is a k-tuple of functions in $\mathcal{S}(R^n)$ that satisfy

$$\sup_{t>0} \sum_{i=1}^{k} |\hat{\phi}^{(i)}(t\xi)| \ge c > 0, \quad \text{all } \xi \in R^n \setminus \{0\}. \tag{1}$$

With the notation

$$F_\phi = \left(F_{\phi^{(1)}}, \dots, F_{\phi^{(k)}} \right), \quad |F_\phi| = \left(\sum_{i=1}^{k} |F_{\phi^{(i)}}|^2 \right)^{1/2},$$

we have

Theorem 2. (a) Let $\phi = (\phi^{(1)}, \dots, \phi^{(k)})$, $k \ge 1$, be a k-tuple of Schwartz functions that satisfies (1) above and suppose that $\psi \in \mathcal{S}$ is such that $\hat{\psi}(\xi) = \sum_{i=1}^{k} \hat{\tau}(\xi) \hat{\phi}^{(i)}(\xi)$ near the origin for some $\hat{\tau} \in C^\infty(R^n)$. Then, for every $a, N > 0$, there is a constant C depending only on ϕ, ψ, p, a, N such that for all tempered distributions f and each (x, t) in R^{n+1}_+ we have

$$|F_\psi(x,t)|^p \le \int_0^{at} \int_{R^n} |F_\phi(y,s)|^p (1 + (|x-y|/s))^{-Np} s^{-n} (s/t)^{Np} dy \frac{ds}{s}$$
$$+ C \int_{R^n} |F_\phi(y,t)|^p (1 + (|x-y|/t))^{-Np} dy.$$

In fact, if $\hat{\psi}$ vanishes in a neighbourhood of the origin, the second integral above may be ommited.

(b)   If now the assumption on ψ is replaced by the weaker condition: All Taylor polynomials of $\hat{\psi}$ at the origin are contained in the ideal of polynomials generated by the Taylor polynomials of $\hat{\phi}$ at the origin, then there is a constant C which depends only on ϕ, ψ, p, a, N such that for all tempered distributions f and (x,t) in R_+^{n+1},

$$|F_\psi(x,t)|^p \le C \int_0^\infty \int_{R^n} |F_\phi(y,s)|^p (1+(|x-y|/s))^{-Np} s^{-n} \min(s/t, t/s)^{Np} \, dy \, \frac{ds}{s}.$$

The following result is also important in applications, cf. Chapter VI.

Theorem 3.   Let ϕ, ψ be Schwartz functions on R^n, suppose that ϕ has a non-vanishing integral, and let Γ denote the cone $\Gamma = \{\xi \in R^n : \xi = (\xi_1, \ldots, \xi_n) = (\xi_1, \xi') \text{ and } |\xi'| < c\xi_1\}$. Then, for all p, $0 < p < \infty$, and each $N > 0$ there is a constant C depending only on ϕ, ψ, p, N, such that if $f \in S'(R^n)$ has support contained in Γ, then for $(x,t) \in R_+^{n+1}$ we have

$$|F_\psi(x,t)|^p \le C \liminf_{\varepsilon \to 0} \int_{R^n} |F_\phi(y,\varepsilon)|^p (1+(|x-y|/\varepsilon))^{-Np} \varepsilon^{-n} dy.$$

In applications we also often deal with functions $f \in \hat{\mathcal{D}}_0$; in that case we have

Theorem 4.   Suppose $\phi = (\phi^{(1)}, \ldots, \phi^{(k)})$ is a k vector of Schwartz functions that satisfy condition (1), and let $\psi \in S'(R^n)$ be a tempered distribution such that $D^\alpha \hat{\psi}$ coincides locally in $R^n \setminus \{0\}$ with an $L^q_{loc}(R^n)$ function, $1 \le q \le 2$, for $|\alpha| \le m$, and

$$\left(\frac{1}{r^n} \int_{\{r < |\xi| < 2r\}} |D^\alpha \hat{\psi}(\xi)|^q \, d\xi \right)^{1/q} \le c_\alpha \frac{1}{r^{|\alpha|}} \min(r^A, r^{-B}),$$

for all $r > 0$, $|\alpha| \le m$. Then, for all p, $0 < p < \infty$, $0 \le A' < A$, $0 \le B' < B$ and $m' \le m - n(\frac{1}{q} - \frac{1}{p})_+$, there is a constant C depending only on $\phi, \psi, p, A, A', B, B', m, m'$, such that for any $f \in \hat{\mathcal{D}}_0$ and $(x,t) \in R_+^{n+1}$, $|F_\psi(x,t)|^p$ is dominated by

$$C \int_0^\infty \int_{R^n} |F_\phi(y,s)|^p (1+(|x-y|/s))^{-m'} s^{-n} \min((t/s)^{A'}, (s/t)^{B'})^p \, dy \, \frac{ds}{s}.$$

Note that we can take $m' = m$ in the above inequality provided that $p \le q$. Also, we could of course allow the range $2 < q \le \infty$ and use Hölder's inequality in the condition on $D^\alpha \hat{\psi}$ to reduce it to the case $q = 2$. This would then require that $p \le 2$ and that $m' < m - n(1/q - 1/p)$.

Remarks. Theorems 1-4 admit a formulation with weights. Namely, if we replace the Lebesgue measure dy in the integrals above by a weighted measure $d\nu(y) = w(y)dy$ with respect to the Lebesgue measure and replace the left-hand side by $|F_\psi(x,t)|^p \nu(B(x,t))t^{-n}$, then Theorems 1,2 and 4 are true provided that ν is doubling, and Theorem 3 is true if $w \in A_\infty$. Of course the constant C now depends on w as well.

Also, the integral with respect to the measure ds/s in Theorems 1,2 and 4 can be replaced by a sum at the points $s_i = \gamma \beta^i t$ provided $\beta < 1$ is close enough to 1 (the choice depending on ϕ), and γ is any positive number in the interval $(\beta, 1]$, except that in Theorem 2 we must pick $\gamma = 1$. The index i in the sum runs over those integers such that the s_i's lie in an interval corresponding to the integration interval of the measure ds/s in those theorems. We refer to this instance as the "discrete version", while the original formulation is referred to as the "continuous version."

Finally, the important particular instance of the Poisson kernel

$$P(x) = c_n(1 + |x|^2)^{(n+1)/2}$$

in place of $\phi \in \mathcal{S}(R^n)$ in Theorems 1,2 and 4 holds as well provided that the extension $F_P(x,t)$ is well defined. As for Theorem 3, it is also true if we replace $\liminf$ by $\limsup$ in the conclusion. Moreover, if $\nabla P(x)$ denotes the (space) gradient of P,

$$\nabla P(x) = (n+1)c_n(1 + |x|^2)^{-(n-1)/2}(x_1,\dots,x_n),$$

note that we may also replace ϕ by ∇P in Theorems 2 and 4. Also, if $\psi = \frac{\partial}{\partial x_i}P$, $i = 1,\dots,n$, and $\psi = P - \eta$ with $\eta \in \mathcal{S}$ with integral equal to 1, then ψ satisfies the assumptions of Theorem 4 with $A = 1$ for any $B > 0$ and $m \geq 0$.

Before describing the applications of these results we proceed to prove a somewhat simpler estimate and then, by means of a partition of unity argument in the Fourier transform side (Lemma 6), we carry over the arguments to prove Theorems 1-4.

We begin by proving

Theorem 5. Let $\phi, \psi \in \mathcal{S}(R^n)$, and assume that $\hat\phi$ is compactly supported and that $\hat\psi = \hat\tau\hat\phi$ for some $\tau \in C^\infty(R^n)$. Then for all p, $0 < p < \infty$, all doubling weighted measures $d\nu(y) = w(y)dy$ with respect to the Lebesgue measure on R^n and every $N > 0$ there is a constant C depending only on ϕ, ψ, w, p, N such that for all $f \in \mathcal{S}'(R^n)$ and $(x,t) \in R_+^{n+1}$ we have

$$|f * \psi(x)|^p \nu(B(x,1)) \leq C \int_{R^n} |f * \phi_s(y)|^p (1 + |x - y|)^{-Np} w(y)\, dy.$$

Proof. We may assume that $\hat\tau$ has compact support and consequently, $|\tau(y)| \leq c_N(1 + |y|)^{-N}$ for all $N > 0$. Therefore,

$$|f * \psi(x)| = |\tau * (f * \phi)(x)| \leq c_N \int_{R^n} |f * \phi(y)|(1 + |x - y|)^{-N}\, dy \qquad (2)$$

for all $N > 0$, which gives the result in the unweighted case when $p = 1$. By Hölder's inequality we also get the result, in the unweighted case, for $p > 1$.

When $0 < p < 1$, and in the general case, we first consider $\phi = \psi$. Observe that since $\hat{\phi}$ has compact support we can write $\hat{\phi} = \hat{\tau}\hat{\phi}$ with $\tau \in \mathcal{D}$. Whence from (2) it follows that

$$|f * \phi(x)| \le c_N \int_{R^n} |f * \phi(y)|(1 + |x - y|)^{-N} \, dy, \quad N > 0. \tag{3}$$

Now, for each $N > 0$ let

$$M_{N,\phi}(f, x) = M_N(x) = \sup_y |f * \phi(y)|(1 + |x - y|)^{-N};$$

for each fixed N, $M_N(x)$ is finite for all x or identically infinite. Assume that the former occurs, and note that

$$|f * \phi(y)| \le (1 + |x - y|)^N M_N(x), \quad \text{all } x, y \text{ in } R^n.$$

Let $\eta(x, y, z) = (1 + |x - y|)/((1 + |x - z|)(1 + |y - z|))$. Since $1 + |x - y| \le 1 + |x - z| + |y - z| \le (1 + |x - z|)(1 + |y - z|)$, we get that

$$\sup_z \eta(x, y, z) \le 1. \tag{4}$$

Now, from (3) it follows that when $0 < p < 1$,

$$|f * \phi(z)| \le c_N \int_{R^n} |f * \phi(y)|^p (1 + |x - y|)^{N(1-p)} M_N(x)^{1-p}(1 + |y - z|)^{-N} \, dy.$$

Whence, multiplying through by $(1 + |x - y|)^{-N}$, taking the sup over z, and using (4), we get

$$M_N(x) \le c_N M_N(x)^{1-p} \int_{R^n} |f * \phi(y)|^p (1 + |x - y|)^{-Np}(\sup_z \eta(x, y, z))^N \, dy$$

$$\le c_N M_N(x)^{1-p} \int_{R^n} |f * \phi(y)|^p (1 + |x - y|)^{-Np} \, dy.$$

Thus,

$$|f * \phi(x)|^p \le M_N(x)^p \le c_N \int_{R^n} |f * \phi(y)|^p (1 + |x - y|)^{-Np} \, dy, \tag{5}$$

provided of course that $M_N(x)$ is finite. Now, since f is a tempered distribution and $\phi \in \mathcal{S}(R^n)$, $f * \phi(x)$ is dominated by a finite number of the Schwartz norms of $\phi(x - \cdot)$, and consequently, $M_N(x)$ is finite for all $N \ge N_f$, where N_f is a large integer that depends on f. Also, since the factor $(1 + |x - y|)^{-Np}$ decreases as N increases, from (5) we get

$$|f * \phi(x)|^p \le c_{N,f} \int_{R^n} (1 + |x - y|)^{-Np} \, dy, \quad x \in R^n, \tag{6}$$

with a constant $c_{N,f}$ that depends on N, f, but not on x. From (6) it follows that $M_N(x)$ is finite, for

$$(|f * \phi(z)|(1 + |x - z|)^{-N})^p$$
$$\le c_{N,f} \int_{R^n} |f * \phi(y)|^p (1 + |x - z|)(1 + |y - z|)^{-Np} \, dy$$
$$\le c_{N,f} \int_{R^n} |f * \phi(y)|^p (1 + |x - y|)^{-Np} \, dy.$$

What this implies is that we obtain (5) for all $N > 0$ provided the right-hand side of (5) is finite; if it is infinite there is nothing to prove.

Note that the constant c_N in (5) is independent of f. Now, for an arbitrary $\psi \in S(R^n)$, from (5), (2) and the definition of $M_N(x)$ we get

$$|f * \psi(x)| \le c_{N'} \int_{R^n} M_N(x)(1 + |x - y|)^{N - N'} \, dy \le c c_{N'} M_N(x)$$
$$\le c c_{N'} c_N^{1/p} \left(\int_{R^n} |f * \phi(y)|^p (1 + |x - y|)^{-Np} \, dy \right)^{1/p},$$

provided we pick $N' > N + n$. This completes the proof in the unweighted case.

As for the weighted case, note that, similarly to (3), we have

$$|\nabla(f * \phi)(x)| \le c_N \int_{R^n} |f * \phi(y)|(1 + |x - y|)^{-n} \, dy \tag{7}$$

which is also a direct consequence of (2) since we can write $(\partial \phi / \partial x_i)(x)\hat{\ } = \hat{\tau}_i \hat{\phi}$, $i = 1, \ldots, n$ with $\tau_i \in \hat{\mathcal{D}}$. Let $M_N(x)$ be defined as before, and set

$$m_{N,\phi}(f, x) = m_N(x) = \sup_y |\nabla(f * \phi)(y)|(1 + |x - y|)^{-N}.$$

From (7) with $N' > N$ we get

$$|\nabla(f * \phi)(z)|(1 + |x - z|)^{-N}$$
$$\le c_{N'} \int_{R^n} M_N(x)((1 + |x - y|)/(1 + |x - z|))^N (1 + |z - y|)^{-N'} \, dy$$
$$\le c_{N'} M_N(x) \int_{R^n} (1 + |z - y|)^{N - N'} \, dy \le c c_{N'} M_N(x).$$

Thus, $m_N(x) \leq c_N M_N(x)$. Assume now that $M_N(x)$ is finite and let x_0 be such that

$$|f * \phi(x_0)| \geq \frac{1}{2} M_N(x)(1 + |x - x_0|)^N.$$

If $|y - x_0| < 1$ we see that

$$|\nabla(f * \phi)(y)| \leq m_N(x)(1 + |x - y|)^N \leq c_N M_N(x)(1 + |x - x_0|)^N$$
$$\leq 2^N c_N M_N(x)(1 + |x - x_0|)^N.$$

Thus there is a ball $B(x_0, r_N)$, $0 < r_n < 1$ such that

$$|f * \phi(y)| \geq \frac{1}{4} M_N(x)(1 + |x - x_0|)^N, \quad y \in B(x_0, r_N),$$

and consequently, it follows that

$$\int_{R^n} |f * \phi(y)|^p (1 + |x - x_0|)^{Np} (1 + |x - y|)^{-N_0 p} w(y) \, dy$$

$$\geq (M_N(x)/4)^p \int_{B(x_0, r_N)} (1 + |x - x_0|)^{Np} (1 + |x - y|)^{-N_0 p} w(y) \, dy$$

$$\geq c M_N(x)^p (1 + |x - x_0|)^{(N - N_0)p} \nu(B(x_0, r_N))$$
$$\times \nu(B(x, 1)) / \nu(B(x_0, 1 + |x - x_0|)),$$

where in the last step we used that $B(x, 1) \subseteq B(x_0, 1 + |x - x_0|)$. Since $\nu \in D_b$ for some b, we have

$$\nu(B(x_0, r_N)) / \nu(B(x_0, 1 + |x - x_0|)) \geq c(1 + |x - x_0|)^{-b}.$$

Hence if $N_0 \leq N - b/p$ we get

$$|f * \phi(x)|^p \leq M_N(x)^p$$
$$\leq c_N \frac{1}{\nu(B(x, 1))} \int_{R^n} |f * \phi(y)|^p (1 + |x - y|)^{-N_0 p} w(y) \, dy, \quad (8)$$

provided $M_N(x)$ is finite. As before this estimate gives

$$|f * \phi(x)|^p \leq c_{N_0, \nu} \frac{1}{\nu(B(x, 1))} \int_{R^n} |f * \phi(y)|^p (1 + |x - y|)^{-N_0 p} w(y) \, dy,$$

for $x \in R^n$. Whence, if $N = N_0 b/p$, it follows that $(|f * \phi(z)|(1 + |x - z|)^{-N})^p$ is dominated by

$$c_{N_0, f}(1 + |x - z|)^{Nq} \nu(B(z, 1))$$

$$\times \int_{R^n} |f * \phi(y)|^p ((1 + |z - y|)(1 + |x - z|))^{-N_0 p} w(y) \, dy$$

$$\leq c_{N_0, f} \frac{1}{\nu(B(x, 1))} \int_{R^n} |f * \phi(y)|^p (1 + |x - y|)^{-N_0 p} w(y) \, dy.$$

We conclude then that $M_N(x)$ is finite whenever the left-hand side of (8) is finite and $N = N_0 + b/p$. Thus (8) holds with $N = N_0 + b/p$ for all $N_0 > 0$, and the constant c_N is independent of f. The estimates for an arbitrary $\psi \in \mathcal{S}(R^n)$ can be obtained exactly as in the unweighted case ($p < 1$), and is therefore left to the reader. ∎

Proof of Theorems 1, 2 and 4. The proof of Theorem 5 was based on the simple identity $\check{\psi} = \hat{\tau}\hat{\phi}$. Now we will use a more complicated identity involving $(\phi_t)\hat{~}(\xi) = \hat{\phi}(t\xi)$ for different values of t. With this identity, given in Lemma 6 below, we are able to get an estimate that corresponds to (2), but with the added ingredient that it involves a sum, or an integral, over different t values. As we will see, an argument analogous to that in the proof of Theorem 5 can then be used to obtain the estimates for the different values of p and the weighted measures.

Lemma 6. Suppose $\phi = (\phi^{(1)}, \ldots, \phi^{(k)})$, $k \geq 1$, is a k-tuple of Schwartz functions which satisfies condition (1) above. Then, there are numbers β_0, R_1, R_2, $0 < \beta_0 < 1$, $0 < R_1 < R_2$, such that for every $\beta \in [\beta_0, 1)$, there is a k-tuple $\eta = (\eta^{(1)}, \ldots, \eta^{(k)})$ of Schwartz functions with $\operatorname{supp} \hat{\eta}^{(i)} \subseteq \{\xi : R_1 < |\xi| < R_2\}$, $1 \leq i \leq k$, and

$$\sum_{j=1}^{\infty} \sum_{i=1}^{k} \hat{\phi}^{(i)}(\beta^j \xi) \hat{\eta}^{(i)}(\beta^j \xi) = 1, \quad \xi \neq 0.$$

Furthermore, the Schwartz norms of $\eta^{(i)}$, $1 \leq i \leq k$, are uniformly bounded in β when β lies in an interval $[\beta_0, c] \subset [\beta_0, 1)$.

Proof. First a remark. By multiplying η by a constant once the lemma is established, also the continuous version, to wit,

$$\int_0^{\infty} \sum_{i=1}^{k} \hat{\phi}^{(i)}(t\xi) \hat{\eta}^{(i)}(t\xi) \frac{dt}{t}, \quad \xi \neq 0,$$

is readily seen to hold. Also, if $k = 1$ and ϕ has nonvanishing integral, β_0 and R_2 can be chosen arbitrarily small.

Let $S^{n-1} = \{\xi \in R^n : |\xi| = 1\}$ denote the unit sphere in R^n. By (1) and the continuity of $\hat{\phi}$, to each $\xi_0 \in S^{n-1}$ we can assign an interval I_{ξ_0} in the line and a number r_0 such that

$$\sum_{i=1}^{k} |\hat{\phi}^{(i)}(t\xi)|^2 \geq c/2 > 0, \quad \text{all } t \in I_\xi \text{ and } \xi \in B(\xi_0, r_0) \cap S^{n-1}.$$

Since S^{n-1} can be covered with a finite number of these neighborhoods, say N, there is a finite collection of intervals $\{I_{\xi_j}\}_{j=1}^{N}$ such that

$$\max_j \inf_{t \in I_{\xi_j}} \sum_{i=1}^{k} |\hat{\phi}^{(i)}(t\xi)|^2 \geq c/2 > 0, \quad \text{all } \xi \in S^{n-1}.$$

Also, there is a β_0, $0 < \beta_0 < 1$, such that to each β, $\beta_0 \le \beta < 1$, $t > 0$ and $j = 1, \ldots, N$, there corresponds an integer h such that $\beta^j t \in I_{\xi_h}$. Let $R_2 > R_1 > 0$ be such that $\bigcup_{j=1}^N I_{\xi_j} \subseteq [2R_1, R_2/2]$, and let $\sigma \in \mathcal{D}$, $\sigma \ge 0$, be such that $\sigma = 1$ on $[2R_1, R_2/2]$ and vanishes off $[R_1, R_2]$. Then we get

$$\sum_{j=-\infty}^{\infty} \sigma(\beta^j|\xi|) \sum_{i=1}^{k} |\hat{\phi}^{(i)}(\beta^i\xi)|^2 = \varrho(\xi) \ge c > 0.$$

It is readily seen that $\varrho \in C^\infty(R^n)$ is homogeneous of degree zero, i.e., $\varrho(\beta\xi) = \varrho(\xi)$. Set now $\hat{\eta}^{(i)}(\xi) = \sigma(\xi)\hat{\phi}^{(i)}(\xi)/\varrho(\xi)$, $1 \le i \le k$. We leave to the reader to verify that the k-tuple $\eta = (\eta^{(1)}, \ldots, \eta^{(k)})$ satisfies the desired properties. ∎

Proof of Theorem 1. By the continuous version of Lemma 6 we see that

$$\hat{\zeta}(\xi) = 1 - \int_0^a \hat{\phi}(s\xi)\hat{\eta}(s\xi)\frac{ds}{s}, \quad 0 < a < \infty,$$

is a function in $\mathcal{D}$; also, by letting R_2 in that lemma be small enough, we get that $\operatorname{supp}\hat{\zeta} \subseteq \{\xi : |\hat{\phi}(a\xi/2)| \ge c\}$ for some $c > 0$. It also follows that $\hat{\zeta}(\xi)/\hat{\phi}(r\xi) \in \mathcal{D}$ for any $r \in [a/2, a]$. Let $\vartheta \in C_0^\infty([a/2, a])$ be a nonnegative function so that $\int_0^\infty \vartheta(s) \, ds/s = 1$, and set

$$\hat{\tau}(s\xi, s) = \hat{\psi}(\xi)\left(\hat{\eta}(s\xi) + \vartheta(s)\hat{\zeta}(\,xi)/\hat{\phi}(s\xi)\right). \tag{9}$$

Then we have the identity

$$\hat{\psi}(\xi) = \int_0^a \hat{\tau}(s\xi, s)\hat{\phi}(s\xi)\frac{ds}{s}, \tag{10}$$

which is the substitute to the identity $\hat{\psi} = \hat{\tau}\hat{\phi}$ in Theorem 5. Then, for every $0 < s \le a$ and a multi-index α we have

$$\begin{aligned}|D^\alpha\hat{\tau}(\xi, s)| &\le |D^\alpha(\hat{\psi}(\xi/s)\hat{\eta}(\xi))| + \vartheta(s)|D^\alpha(\hat{\zeta}(\xi/s)/\hat{\phi}(\xi))| \\ &\le c_{\alpha,N}(1 + (|\xi|/s))^{-N}s^{-|\alpha|}\chi_{[R_1,R_2]}(|\xi|) + c_\alpha\vartheta(s)\chi_{[0,2R_2]}(|\xi|) \\ &\le c_{\alpha,N}s^{N-|\alpha|}\chi_{[0,2R_2]}(|\xi|),\end{aligned}$$

for all $N > 0$. It thus follows that the inverse Fourier transform of $\hat{\tau}(s\xi, s)$ satisfies the estimate

$$|\tau_s(x, s)| \le c_N s^N(1 + (|x|/s))^{-N}s^{-n}, \quad x \in R^n, \ 0 < s \le a.$$

From (10) we get

$$F_\psi(x,t) = \int_0^a (\tau_{ts}(\cdot,s) * F_\phi(\cdot,ts))(x)\,\frac{ds}{s}$$

$$= \int_0^{at} (\tau_s(\cdot,s/t) * F_\phi(\cdot,s))(x)\,\frac{ds}{s},$$

and consequently, it follows that for every $N > 0$,

$$|F_\psi(x,t)| \le c_N \int_0^{at} \int_{R^n} |F_\phi(y,s)|(1+(|x-y|/s))^{-N}s^{-n}(s/t)^N\,dy\frac{ds}{s}. \qquad (11)$$

Since for $0 < s \le at$ we have

$$\frac{1}{c}(1+(|x-y|/s))^{-2N}s^{2N} \le (1+(|x-y|/t))^{-2N}s^{2N}$$

$$\le c(1+(|x-y|/s))^{-N}s^{-N},$$

we can replace $(1+(|x-y|/s))^{-N}$ by $(1+(|x-y|/t))^{-N}$ in Theorem 1, and instead of (11) above we write, with $a = 1$ there,

$$|F_\psi(x,t)| \le c_N \int_0^t \int_{R^n} |F_\phi(y,s)|(1+(|x-y|/t))^{-N}s^{-n}(s/t)^N\,dy\frac{ds}{s}. \qquad (12)$$

The proof proceeds now as that of Theorem 5, but with (12) instead of (2) in that theorem. Let

$$M_N(x,t) = \sup_{y\in R^n, s\le t} |F_\phi(y,s)|(1+(|x-y|/t))^{-N}(s/t)^N.$$

By using the elementary inequality, which serves as a substitute for (4) in Theorem 5,

$$\frac{t}{s}(1+(|x-y|/t)) \le \frac{t}{r}(1+(|x-z|/t))\frac{r}{s}(1+(|y-z|/r)),$$

for x,y,z in R^n and $0 < s < r < t$, we obtain the estimate of Theorem 1, in the unweighted case, for all $0 < p \le 1$.

As for the weighted case, let

$$m_N(x,t) = \sup_{y\in R^n, s\le t} |s\nabla F_\phi(y,s)|(1+(|x-y|/t))^{-N}(s/t)^{-N},$$

and observe that from (12) it follows that $m_N(x,t) \le cM_N(x,t)$. Now, if $M_N(x,t)$ is finite, we can find a ball $B((x_0,t_0),r_Nt_0)$ in R_+^{n+1} where $|F_\phi(x,t)| \ge \frac{1}{4}(1+(|x_0-x|/t))^N(t_0/t)^{-N}$, and the weighted version of Theorem 1 with $\phi = \psi$

holds. We may now combine this estimate with (11) to obtain the estimate for arbitrary ψ, and to complete the proof. ∎

Proof of Theorem 2.(a)   From the discrete version of Lemma 6 we get that if $a > 0$,

$$\hat{\zeta}(\xi) = 1 - \sum_{j, \beta^j \le a} \langle \hat{\phi}(\beta^j \xi), \hat{\eta}(\beta^j \xi) \rangle$$

belongs to $\mathcal{D}$, and if a is large enough, then $\operatorname{supp} \hat{\zeta} \subseteq \{\xi : |\xi| \le ca\}$, where the identity $\hat{\psi} = \langle \hat{\tau}, \hat{\phi} \rangle$ holds. Whence,

$$\hat{\psi}(\xi) = \sum_{j, \beta^j \le a} \langle \hat{\tau}(\beta^j \xi, j), \hat{\phi}(\beta^j \xi) \rangle + \hat{\zeta}(\xi) \langle \hat{\tau}(\xi), \hat{\phi}(\xi) \rangle$$

where for $i = 1, \ldots, k$,

$$\hat{\tau}(\xi, j) = \left(\hat{\tau}^{(1)}(\xi, j), \ldots, \hat{\tau}^{(k)}(\xi, j) \right), \quad \text{and} \quad \hat{\tau}^{(i)}(\xi, j) = \hat{\psi}(\beta^{-j} \xi) \hat{\tau}^{(i)}(\xi).$$

By estimating the Schwartz norms as before we get

$$|F_\psi(x, t)| \le c_N \sum_{j, \beta^j \le a} \beta^{jN} \int_{R^n} |F_\phi(y, \beta^j t)| (1 + (|x - y|/t))^{-N} (\beta^j t)^{-n} \, dy$$

$$+ c_N \int_{R^n} |F_\phi(y, t)| (1 + (|x - y|/t))^{-N} t^{-n} \, dy.$$

Now we define for $N > 0$

$$M_N(x, t) = \sup_{\substack{y \in R^n \\ j, \beta^j \le a}} |F_\phi(y, \beta^j t)| (1 + (|x - y|/t))^{-N} \beta^{jN},$$

and

$$m_N(x, t) = \sup_{\substack{y \in R^n \\ j, \beta^j \le a}} |(\beta^j t) \nabla F_\phi(y, \beta^j t)| (1 + (|x - y|/t))^{-N} \beta^{jN}.$$

Here, as usual, ∇F_ϕ denotes the gradient in the space variables alone, and

$$|\nabla F_\phi| = \sum_{i=1}^{k} \sum_{j=1}^{n} \left| \frac{\partial}{\partial x_j} F_\phi^{(i)} \right|.$$

Note that locally we can write $\left(\frac{\partial}{\partial x_j} \right)\hat{} = \hat{\tau}_j \hat{\phi}^{(i)}$ with $\hat{\tau}_j \in C^\infty(R^n)$. Thus, we can use estimate (13) with $t|\nabla F(x, t)|$ on the left-hand side, and with $a > 0$

arbitrarily small. Whence, it follows that $m_N(x,t) \leq c_N M_N(x,t)$, and with an argument similar to that in the proof of Theorem 1, we note that

$$|F_\psi(x,t)^p| \leq c M_N(x)^p$$

$$\leq \frac{c_N}{\nu(B(x,t))} \left(\sum_{j,\beta^j \leq a} \int_{R^n} |F_\phi(y,\beta^j t)|^p (1 + (|x - y|/t))^{-Np} \beta^{jNp} \frac{w(y)}{t^n} \, dy \right.$$

$$\left. + t^{-n} \int_{R^n} |F_\phi(y,t)|^p (1 + |x - y|)^{-Np} w(y) \, dy \right).$$

Since the Schwartz norms of the $\eta^{(i)}$'s in Lemma 6 are uniformly bounded in β in the interval $[\beta_0, \beta_0^{1/2}]$, we can find a constant c_N such that the above estimate holds for all $\beta \in [\beta_0, \beta_0^{1/2}]$. By integrating over β in this interval, the continuous version of Theorem 2(a) obtains.

As for the proof of part (b), we need the following result.

Lemma 7. Let ϕ and ψ be as in the hypothesis of Theorem 2 (b), and let N be a positive integer. Then there is a bounded vector-valued function h_N with compact support on R_+^{n+1} such that the function ψ_N defined by

$$\psi_N(x) = \int \int_{R_+^{n+1}} \langle \phi_s(x - y), h_N(y,s) \rangle \, dy \, \frac{ds}{s}$$

belongs to $\mathcal{S}(R^n)$, and

$$|\hat{\psi}(\xi) - \hat{\psi}_N(\xi)| \leq c|\xi|^{N+1}, \quad \text{near the origin.}$$

Proof. Let $\omega_0, \omega_1, \ldots, \omega_N$ denote the dual basis to $1, t, \ldots, t^N$ in the subspace of polynomials of degree less that or equal to N in $L^2([1,2])$, i.e., for $0 \leq j, k \leq N$ we have

$$\int_1^2 \omega_j(t) t^k \, dt = \begin{cases} 1 & \text{whenever } j = k \\ 0 & \text{otherwise.} \end{cases}$$

Let $\sum_j P_j(\xi)$ and $\sum_j Q_j(\xi)$ denote the Taylor series at the origin of $\hat{\phi}$ and $\hat{\psi}$ respectively, where $P_j = (P_{j1}, \ldots, P_{jk})$ and similarly Q_j are homogeneous polynomials of degree j. Then,

$$\int_1^2 \hat{\phi}_{1/t}(\xi) t^{-n} \omega_j(t) \, dt = P_j(\xi) + E(\xi),$$

where the error term $E(\xi)$ satisfies $|E(\xi)| \leq c|\xi|^{N+1}$ for ξ near the origin. According to our assumptions there are vector-valued polynomials $q_i = (q_{i1}, \ldots, q_{ik})$,

$i = 1, 2, \ldots, N$, such that the Taylor expansion of ψ of order N at the origin can be written

$$\sum_{j=0}^{N} Q_j(\xi) = \sum_{j=0}^{N} \langle q_j(\xi), P_j(\xi) \rangle \,.$$

Let q_j, $j = 0, 1, \ldots, N$ be Schwartz functions with compact support such that $\hat{g}_j(\xi) = q_j(\xi) + E_j(\xi)$, where the error term $|E_j(\xi)| \leq c|\xi|^{N+1}$ for ξ near the origin. Using ω_j and g_j we can build up the desired functions h_N. Indeed, we set

$$h_N(y, s) = \sum_{j=0}^{N} g_j(y) \omega_j(1/s) \chi_{[1/2,1]}(s) s^{n-1} \,.$$

The reader will note that the desired properties of h_N are readily verified. Also, the construction can be carried out so that h_N has support contained in any prescribed compact subset of R_+^{n+1} with nonempty interior. There is also a discrete version of the result where the integration over s is replaced by a summation over $N + 1$ points $s_i = \beta^i$, $i = 0, 1, \ldots, N$. ∎

Proof of Theorem 2 (b). We only consider the case $p = 1$ since the rest of the proof follows along similar lines; we do the continuous version.

First we construct a ψ_N according to Lemma 7. Since the function h_N is bounded and has compact support, it is clear that the desired estimate holds with $|F_\psi(x, t)|$ on the left-hand side. By replacing ψ with $\psi - \psi_N$ we may assume that $\hat{\psi}$ vanishes of order N at the origin; this can be done for any positive integer N. We write

$$\hat{\zeta}(\xi) = 1 - \int_0^1 \hat{\phi}(s\xi) \hat{\eta}(s\xi) \frac{ds}{s} \,.$$

The only difference in the argument is how to handle $F_{\zeta_r * \psi}(x, t)$. Since $\hat{f}$ is also a tempered distribution and it has finite order, it follows that $F_{\zeta_r * \psi}(x, t)$ goes to 0 as r tends to infinity, provided ψ vanishes of order N with $N = N_f$ large enough. From this remark and the above arguments teh desired estimate follows. Unfortunately, the constant C depends now on $N = N_f$, and it therefore depends on f. In order to complete the proof it suffices to prove the following lemma.

Lemma 8. Assume that the right-hand side of the inequality in Theorem 2 (b) is finite for $N = N_0$. Then, for $N = N_0$, $|F_{\zeta_r * (\psi - \psi_N)}|$ tends to 0 as $r \to \infty$.

Proof. We use the estimate in Theorem 2 (b) for $N = N_f$ in the right-hand side of that estimate when $N = N_0$, which is finte. Thus,

$$|F_\psi(x, t)| \leq c_f(x)|t|^{N_0} \,, \quad \text{large } t \,.$$

By the definition of ψ_{N_0} as above we get

$$|F_{\psi_{N_0}}(x,t)| \leq c_f(x)|t|^{N_0}, \quad \text{large } t.$$

By writing $1 - \hat{\zeta}(\xi)$ as an integral it follows that

$$|F_{(\psi-\psi_{N_0})-\zeta_r*(\psi-\psi_{N_0})}(x,t)| \leq c_f(x)|t|^{N_0}, \quad \text{large } t,$$

and consequently, the left-hand side converges as $r \to \infty$. We thus conclude that

$$|F_{\zeta_r*(\psi-\psi_{N_0})}(x,t)| \leq c_f(x)|t|^{N_0}$$

and that $F_{\zeta_r*(\psi-\psi_{N_0})}$ converges as $r \to \infty$. We write

$$\hat{\psi}(t\xi) - \hat{\psi}_{N_0}(t\xi) = \sum_{k=N_0+1}^{N_f} t^k Q_k(\xi) + E(\xi),$$

where $Q_k(\xi)$ is a homogeneous polynomial of degree k, and the error term satisfies $|E(\xi)| \leq c_f(t)|\xi|^{N_f}$ for ξ near the origin. Since N_f was chosen so large that $\hat{f}(\hat{\zeta}(r\cdot)Ee^{ix\cdot\xi})$ goes to 0 as $r \to \infty$, it follows that

$$F_{\zeta_r*(\psi-\psi_{N_0})}(x,t) = \sum_{\ell=N_0+1}^{N_f} t^\ell a_\ell(x,r) + E_1(x),$$

where $E_1(x) \to 0$ as $r \to \infty$. Since the left-hand side of the above equality converges as $r \to \infty$, the limit must be a polynomial in t,

$$\sum_{\ell=N_0+1}^{N_f} t^\ell a_\ell(x),$$

say. Since for large t by the above estimates this polynomial is dominated by $c_f(x)|t|^{N_0}$, it is the zero polynomial. This completes the proof of the lemma and that of Theorem 2 (b). ∎

Proof of Theorem 4. By Lemma 6 we have the identity

$$\hat{f}(\xi)\hat{\psi}(t\xi) = \sum_{j=-\infty}^{\infty} \hat{\psi}(t\xi)\langle\hat{\eta}(\gamma\beta^j t\xi), \hat{\phi}(\gamma\beta^j t\xi)\rangle\hat{f}(\xi)$$

for all $\xi \in R^n$, $t > 0$ and $f \in \hat{\mathcal{D}}_0$. Here $\beta \in [\beta_0, 1)$ is as in Lemma 6, and γ is any number in the interval $[1/2, 1]$. Taking then the inverse Fourier transform,

$$F_\psi(x,t) = \sum_{j=-\infty}^{\infty} \sum_{i=1}^{k} K^{(i)}_{\gamma\beta^j t, t} * F_{\phi^{(i)}}(\cdot, \gamma\beta^j t)(x),$$

where $K_{s,t}^{(i)}$ is the inverse Fourier transform of $\hat{\eta}^{(i)}(s\xi)\hat{\psi}(t\xi)$. Since $K_{s,t}^{(i)}(x) = s^{-n}K_{1,t/s}^{(i)}(x/s)$, by the Hausdorff-Young inequality we get, with $1 \leq q \leq 2$ and $1/q + 1/q' = 1$,

$$\|(1+|x|)^m K_{1,t}^{(i)}\|_{L^{q'}} \leq c \sum_{|\alpha| \leq m} \|D^\alpha \left(\hat{\eta}^{(i)}\hat{\psi}(t\cdot)\right)\|_{L^q}$$

$$\leq c \sum_{|\alpha| \leq m} \left(\int_{|\xi| \sim 1} |D^\alpha(\hat{\psi}(t\xi)|^q \, d\xi\right)^{1/q}$$

$$= c \sum_{|\alpha| \leq m} t^{|\alpha|} \left(t^{-n} \int_{|\xi| \sim t} |D^\alpha \hat{\psi}(\xi)|^q \, d\xi\right)^{1/q}$$

$$\leq c \min(t^A, t^{-B}).$$

The time has come to introduce some notations that will simplify the formulas that follow. Namely,

$$m(t, A, B) = \min(t^A, t^{-B}), \quad \text{and} \quad \eta(y, N) = (1+|y|)^{-N}.$$

We then have

$$|K_{s,t}^{(i)} * F_{\phi^{(i)}}(\cdot, s)(x)| = \left|\int_{R^n} K_{1,t/s}^{(i)}((x-y)/s) F_{\phi_{(i)}}(y, s) s^{-n} \, dy\right|$$

$$\leq \left(\int_{R^n} |K_{1,t/s}^{(i)}((x-y)/s)|^{q'} \eta_s(x-y, -mq') \, dy\right)^{1/q'}$$

$$\times \left(\int_{R^n} |F_{\phi^{(i)}}(y, s)|^q \eta_s(x-y, mq) \, dy\right)^{1/q} = I \times J,$$

say. Moreover, since $I = \|(1+|x|^m)K_{1,t/s}^{(i)}\|_{L^{q'}}$, it follows that

$$|K_{s,t}^{(i)} * F_{\phi^{(i)}}(\cdot, s)(x)| \leq cm(s/t, A, B) \times J.$$

Also, by Hölder's inequality, the following estimate holds: For any $\varepsilon > 0$ and any nonnegative function g on $(0, \infty)$,

$$\sum_{j=-\infty}^{\infty} g(\gamma\beta^j) \leq c_\varepsilon \left(\sum_{j=-\infty}^{\infty} g(\gamma\beta^j)^q m(\beta, \varepsilon j, \varepsilon j)\right)^{1/q}.$$

We then have

$$|F_\psi(x, t)|^p$$

$$\leq c \sum_{j=-\infty}^{\infty} m(\beta, j(Bq - \varepsilon), j(Aq - \varepsilon)) \int_{R^n} |F_\phi(y, \gamma\beta^j t)|^q \eta_{\beta^j t}(x-y, mq) \, dy.$$

Now, when $p > q$, from Hölder's inequality it follows that

$$
|F_\psi(x,t)|^p \leq c \sum_{j=-\infty}^{\infty} m(\beta, j(Bq - \varepsilon'), j(Aq - \varepsilon))
$$
$$
\times \int_{R^n} |F_\phi(y, \gamma\beta^j t)|^p \eta_{\beta^j t}(x - y, mp - (n + \varepsilon)(1/q - 1/p)) \, dy .
$$

On the other hand, when $p < q$ we invoke Theorem 2 instead, with some large N, and obtain

$$
|F_\psi(x,t)|^p \leq c \Big(\sum_{j=-\infty}^{\infty} m(\beta, j(Bq - \varepsilon), j(Aq - \varepsilon)) \eta_{\beta^j t}(x - z, mq)
$$
$$
\times \Big(\sum_{h=j-h_0}^{\infty} \int_{R^n} |F_\phi(y, \gamma\beta^h t)|^p (1 + (|z - y|/\beta^j t))^{-Np} \beta^{(h-j)Np} \, dy \Big) dz \Big) .
$$

Now we use Minkowski's inequality on $L^{q/p}$-norms for functions of the variables $(z, j) \in R^n \times Z$, and observe that

$$
|F_\psi(x,t)|^p \leq c \sum_{h=-\infty}^{\infty} \int_{R^n} |F_\phi(y, \gamma\beta^h t)|^p G(h, \beta, t, x, y)(\beta t)^{-n} \, dy ,
$$

where

$$
G(h, \beta, t, x, y) = \sum_{j=-\infty}^{h+h_0} m(\beta, j(Bq - \varepsilon), j(Aq - \varepsilon)) g(x, y, \beta, t, j)
$$

and $g(x, y, \beta, t, j)$ equals

$$
\int_{R^n} (1 + (|x - z|/\beta^j t))^{-mq} \big((1 + (|z - y|/\beta^j t))^{-Np} \beta^{(h-j)Np} \big)^{q/p} (\beta^j t)^{-n} \, dz
$$
$$
\leq c\beta^{(h-j)Nq}(1 + (|x - y|/\beta^j t))^{-mq} .
$$

Here we used the elementary estimate

$$
(1 + (|x - z|/\beta^j t))^{-mq}(1 + (|z - y|/\beta^j t))^{-Nq}
$$
$$
\leq c(1 + (|x - y|/\beta^j t))^{-mq}(1 + (|z - y|/\beta^j t))^{-Nq - mq} ,
$$

and the fact that

$$
\int_{R^n} \eta_{\beta^j}(z - y, Nq + mq) \, dz \leq c .
$$

Consequently,

$$G(h,\beta,t,x,y) \le cm(\beta,h(Bq-\varepsilon),h(Aq-\varepsilon))(1+(|x-y|/\beta^h t))^{-mq}\,.$$

Thus, finally we get

$$|F_\psi(x,t)|^p$$

$$\le c \sum_{h=-\infty}^{\infty} m(\beta,h(Bq-\varepsilon'),h(Aq-\varepsilon')) \int_{R^n} |F_\phi(y,s)|^p \eta_{\beta^h t}(x-y,mp)\,dy\,,$$

as we wanted to show.

As for the weighted case, we use (14) and an estimate on some quotients involving ν. Indeed, if $\nu \in D_{b_0} \cap RD_{b_1}$, $0 < b_1 \le \eta \le b_0 < \infty$, then

$$\nu(B(x,t))/t^n/\nu(B(z,r))/r^n$$

$$\le c(1+(|x-y|/\max(t,r)))^{b_0-b_1} \max\left((r/t)^{n-b_1},(r/t)^{-(b_0-n)}\right).$$

A similar argument to the one used in the unweighted case shows that

$$\nu(B(x,t))|F_\psi(x,t)|^p$$

$$\le c_\varepsilon \sum_{h=-\infty}^{\infty} m(\beta,h(Bp-(n-b_1)-\varepsilon),h(Ap-(b_0-n)-\varepsilon))g(h,x,t)\,,$$

where now

$$g(h,x,t) = \int_{R^n} |F_\phi(y,\gamma\beta^h t)|^p (1+(|x-y|/\beta^h t))^{-mp+(n+\varepsilon)(\frac{p}{q}-1)^+}$$

$$\times\, (1+(|x-y|/\max(\beta^j t,t)))^{b_0-b_1}\, w(y)\,dy\,.$$

To get the continuous version of Theorem 4 we only need to integrate the discrete version over γ on an interval, $[1/2,1]$ say. This completes the proof. ∎

Proof of Theorem 3. Assume first that $f \in \mathcal{S}(R^n)$ is such that $\hat{f}$ has support contained in $[0,\infty)$; we claim that for $\phi \in \mathcal{S}(R)$ and $0 < p < \infty$, we have

$$|f * \phi_t(x)|^p \le c_N \int_{R^n} |f(y)|^p (1+(|x-y|))^{-Np} t^{-1}\,dy\,. \tag{15}$$

Indeed, let $z = x + it$ and $P_t(x) = \frac{1}{\pi} t/(t^2+x^2)$, the Poisson kernel. Then $f(z) = (f * P_t)(x)$ is analytic in R_+^2 and the pth power of the modulus of the analytic function $f(z)(z+it_0)^{-N}$ is subharmonic. Thus,

$$|f(it_0)(2it_0)^{-N}|^p \le \frac{1}{\pi} \int_R |f(x)|^p (t_0^2+x^2)^{-Np/2} \frac{t_0}{t_0^2+x^2}\,dx\,,$$

and by a translation argument also

$$|f(x + it_0)|^p \le \frac{2^{Np}}{\pi} \int_R |f(x)|^p (1 + (x/t_0))^{-Np/2-1} t_0^{-1} \, dx \, .$$

To get (15), we take a finite linear combination of Poisson kernels, namely, $\phi_N = \sum_{j=1}^{M(N)} a_{j,N} P_{\beta j}$. Provided that $M(N)$ is large enough we can find coefficients $a_{j,N}$ so that ϕ_N will have as many Schwartz norms bounded as are necessary for the conclusion of Theorem 1 to hold, with ϕ replaced by ϕ_N. Since ϕ_N satisfies (15), we get (15) for any $\phi \in \mathcal{S}(R)$ by integration.

We introduce next a smoothed characteristic function of the cone Γ. Let $\varrho \in C_0^\infty([-1,1])$ be such that $\int_R \varrho = 1$, and set

$$\tau(x) = \int_R \chi_\Gamma(x_1 - y, x_2, \dots, x_n) \varrho(y) \, dy \, .$$

Clearly $\tau(x) = 1$ in a neighborhood of Γ and $\tau(x) = 0$ if $x + (1, 0, \dots, 0)$ is not in Γ. Furthermore, $\tau \in C^\infty(R^n)$, and $|D^\alpha \tau| \le c_\alpha$ for all α.

Let $\phi_1 \in \mathcal{S}(R)$ have a nonvanishing integral. Then,

$$\hat{\phi}(\xi) = \hat{\phi}(\xi_1, \dots, \xi_n) = \tau(\xi) \hat{\phi}_1(\xi_1)$$

is a Schwartz function on R^n and $\hat{\phi}(0) = \tau(0)\hat{\phi}_1(0) = \hat{\phi}_1(0) \ne 0$. From this identity it follows that if $f \in \mathcal{S}(R^n)$ and supp $\hat{f} \subseteq \Gamma$, then

$$\hat{f}(\xi)\hat{\phi}(t\xi) = \hat{f}(\xi)\tau(t\xi)\hat{\phi}_1(t\xi) = \hat{f}(\xi_1, \dots, \xi_n)\hat{\phi}_1(t\xi_1) \, .$$

Thus, by taking the inverse Fourier transform,

$$\int_{R^n} f(y)\phi_t(x - y) \, dy = \int_R f(y_1, x_2, \dots, x_n)(\phi_1)_t(x_1 - y_1) \, dy_1 \, ,$$

where the function $f(\cdot, x_2, \dots, x_n)$ for $x_2, \dots, x_n$ fixed is in $\mathcal{S}(R)$ and its Fourier transform is supported in $[0, \infty)$. Consequently,

$$|F_\phi(x,t)|^p = |f(\cdot, x_2, \dots, x_n) * (\phi_1)_t(x_1)|^p$$
$$\le c \int_R |f(y_1, x_2, \dots, x_n)|^p (1 + (|x_1 - y_1|/t))^{-Np} t^{-1} \, dy_1 \, .$$

Whence, if ϕ is an arbitrary Schwartz function on R^N, by Theorem 1 we get,

$$|F_\psi(x,t)|^p \le c \int_0^t \int_{R^n} \left(\int_R |f(y_1, z_2, \dots, z_n)|^p (1 + (|z_1 - y_1|/r))^{-Np} r^{-1} \, dy_1 \right)$$
$$\times (1 + (|z - x|/r))^{-Np} r^{-n} (r/t)^{Np} \, dz \, dr$$
$$= \int_{R^n} |f(y_1, z_2, \dots, z_n)|^p g(t, y_1, z_2, \dots, z_n, r) \, dy_1 \, dz_2 \dots dz_n \, ,$$

where $g(t, y_1, z_2, \ldots, z_n, r)$ is equal to

$$\int_0^t \int_R (1 + (|z - 1 - y_1|/r))^{-Np} r^{-1} (1 + (|z - x|/r))^{-Np} (r/t)^{Np} r^{-n} \, dz_1 \, dr \,.$$

Now, since

$$(1 + (|z_1 - y_1|/r))(1 + (|z - x|/r))$$

$$\geq (1 + (|z_1 - y|/r)) \sum_{j=1}^n (1 + (|z_j - x_j|/r))$$

$$\geq c(1 + (|y_1 - x_1/r)) \sum_{j=2}^n (1 + (|z_j - x_j|/r))$$

$$\geq c(1 + (|(y_1, z_2, \ldots, z_n) - x|/r)) \,,$$

it follows that $g(t, y_1, z_2, \ldots, z_m, r)$ does not exceed

$$c \int_0^t \int_R (1 + (|z_1 - y_1|/r))^{-n-1} r^{-1} \, dz_1$$

$$\times (1 + (|(y_1, z_2, \ldots, z_n) - x|/r))^{-Np+n+1} r^{-n} (r/t)^{Np} \, dr$$

$$\leq c(1 + (|(y_1, z_2, \ldots, z_n) - x|/t))^{-Np+n+1} t^{-n} \,.$$

This gives, for $N_1 \leq N - n + 1$,

$$|F_\psi(x, t)|^p \leq c \int_R |f(y)|(1 + (|x - y|/t))^{-N_1 p} t^{-n} \, dy \,.$$

As for the general result, we need only use an approximation argument. Let $\sigma \in S(R^n)$ be such that its support is contained in Γ, and $\sigma(0) = 1$. Further, let ϕ be a Schwartz function with nonvanishing integral, and let f be a tempered distribution on R^n with support contained in Γ. Then $f_{s,\epsilon}(x) = f * \phi_\epsilon(x) \sigma(sx)$ is a Schwartz function and its Fourier transform $(\hat{f} \hat{\phi}(\epsilon \cdot) * s^{-n} \hat{\sigma}(\cdot/s))(\xi)$ has support contained in Γ. Thus,

$$|f_{s,\epsilon} * \psi_t(x)|^p \leq c \int_R |f_{s,\epsilon}(y)|^p (1 + (|x - y|/t))^{-Np} t^{-n} \, dy$$

$$= c \int_R |F_\phi(y, \varepsilon)|^p |\sigma(sy)|^p (1 + (|x - y|/t))^{-Np} t^{-n} \, dy \,. \qquad (16)$$

Since $\sigma(0) = 1$ and $|\sigma(x)| \leq c$ for all x, the last integral converges to

$$\int_R |F_\phi(y, \varepsilon)|^p (1 + (|x - y|/t))^{-Np} t^{-n} \, dy \,, \quad \text{as } s \to 0 \,,$$

provided, of course, that this integral is finite.

Also,

$$f_{s,\varepsilon} * \psi_t(x) = \int_{R^n} F_\phi(y,\varepsilon)\sigma(sy)\psi_t(x-y)\,dy$$

converges to $\int_{R^n} F_\phi(y,\varepsilon)\psi_t(x-y)\,dy$ as $s \to 0$, since this last integral is absolutely convergent. Furthermore,

$$(f * \phi_\varepsilon) * \psi_t(x) = (f * \psi_t) * \phi_\varepsilon(x),$$

which converges, since $f * \psi_t \in C^\infty(R^n)$ is dominated by $c_{f,\psi,t}(1+|x|)^N$ for some N, to $f * \psi_t(x)$ as ε tends 0. It then follows that, as asserted,

$$|f * \psi_t(x)|^p \le c \liminf_{\varepsilon \to 0} \int_{R^n} |F_\phi(y,\varepsilon)|^p (1 + (|x-y|/t))^{-Np} t^{-n}\,dy.$$

As for the weighted version, we use Hölder's inequality and the fact, discussed in Chapter I, that

$$\int_{R^n} w(y)^{-1/(p-1)}(1 + (|x-y|/t))^{-N} t^{-n}\,dy \le c\nu(B(x,t))^{-1/(p-1)},$$

whenever $w \in A_p$ and N is sufficiently large. ∎

The first application of these results are estimates between some of the functions introduced in Chapter IV.

Theorem 8. Let ν be a doubling weighted measure with respect to the Lebesgue measure on R^n, $d\nu(x) = w(x)dx$, and suppose ϕ, ψ are Schwartz functions on R^n and that ϕ has nonvanishing integral. Then for a tempered distribution f on R^n and $0 < p < \infty$ we have

$$M_0(F_\psi, x)^p \le cM_\nu(M_0(F_\phi)^p)(x)$$

and

$$M_1(F_\psi, x)^p \le cM_\nu(M_0(F_\phi)^p)(x).$$

Also, provided that $\nu \in D_b$ and $\lambda \ge b/p$,

$$N_\lambda(F_\psi, x)^p \le cM_\nu(M_0(F_\phi)^p)(x).$$

Note that this theorem holds with ϕ, ψ as in Theorem 2 as well.

We need now one more definition, namely,

$$g_p(F,x) = g_p(x) = \left(\int_0^\infty |F(x,t)|^p \frac{dt}{t}\right)^{1/p}, \quad 0 < p < \infty.$$

Theorem 9. Let ϕ and ψ be functions as in Theorem 1 (or as in Theorem 2). Then, if $0 < p_1 \leq p_0 \leq \infty$ and $\lambda_0 \geq \lambda_1 > 0$ we have

$$g^*_{\lambda_0,p_0}(F_\psi, x) \leq cg^*_{\lambda_1,p_1}(F_\phi, x).$$

We also have

$$g_{p_0}(F_\psi, x), S_{1,p_0}(F_\psi, x) \leq cg^*_{\lambda_1,p_1}(F_\phi, x),$$

for all extensions F_ϕ, F_ψ of a tempered distribution f.

Theorem 10. Let ϕ and ψ be as in Theorem 4, with $A, B, m > 0$. If $0 < \lambda < B - (n/p)$ and $\lambda \leq m + (n + \varepsilon)(\frac{1}{p} - \frac{1}{q})^+$ for some $\varepsilon > 0$, then we have

$$g^*_{\lambda_0,p}(F_\psi, x) \leq cg^*_{\lambda,p}(F_\phi, x),$$

and

$$g_p(F_\psi, x), S_{1,p}(F_\psi, x) \leq cg^*_{\lambda,p}(F_\phi, x)$$

for all extensions F_ϕ, F_ψ of a function $f \in \hat{\mathcal{D}}_0$, provided that $\lambda_0 \geq \lambda$ and $\lambda_0 > (n/p) - A$.

A couple of remarks concerning these results. First, since $S_{1,p}(F, x) \leq cg^*_{\lambda,p}(F, x)$ for any F and $\lambda > 0$, and since the weighted norms of $g^*_{\lambda,p}(F, x)$ can be estimated by those of $S_{1,p}(F, x)$ provided that λ is sufficiently large, the exact value depending on w, these norms are actually equivalent. As for the estimates in terms of $g_p(F, x)$, see Theorem 11 below.

Also, let $\Psi = \{\psi\}$ be a family of tempered distributions that satisfies the estimate

$$\left(\frac{1}{r^n} \int_{\{|\xi| \sim r\}} |D^\alpha \psi(\xi)|^q \, d\xi\right) \leq c_\alpha r^{-|\alpha|} \min(r^A, r^{-B}), \quad |\alpha| \leq m,$$

uniformly for $\psi \in \Psi$. If

$$F_\Psi(x, t) = \sup_{\psi \in \Psi} |F_\psi(x, t)|,$$

we can replace F_ψ by F_Ψ in Theorem 10; this will be useful in the treatment of Fourier multipliers in chapter XI.

Theorem 11. Let ν be a doubling weighted measure with respect to the Lebesgue measure on R^n, $d\nu(x) = w(x)dx$, and suppose ϕ, ψ are as in Theorem 2. Then, for all p and q, $0 < p < q < \infty$, we have

$$\|S_{1,p}(F_\psi)\|_{L^q_w} \leq c\|g_p(F_\phi)\|_{L^q_w}.$$

Moreover, if $\lambda > \lambda_0 = \lambda_0(p,q,w)$ (this is the same value as the λ_0 in Theorem 3 of Chapter IV), we also have

$$\|g^*_{\lambda,p}(F_\psi)\|_{L^p_w} \leq c\|g_p(F_\phi)\|_{L^q_w}.$$

Before we sketch the proofs of these results we discuss the convergence of the extension $F_\phi(x,t)$ as $t \to 0$. We say that F_ϕ converges nontangentially to L at x_0, if for every $r > 0$

$$\lim_{\epsilon \to 0} \sup_{|y-x_0|<rt<\epsilon} |F_\phi(y,t) - L| = 0.$$

Analogously, we say that F_ϕ converges radially to L at x_0, if

$$\lim_{t \to 0} |F_\phi(x_0,t) - L| = 0.$$

We then have the following results.

Theorem 12. Let $0 < p < \infty$, $\lambda > 0$, and suppose $\phi \in S(R^n)$. If x_0 in R^n is such that $g^*_{\lambda,p}(F_\phi, x_0) < \infty$, then $F_\phi(x,t)$ converges nontangentially to 0 at x_0.

Theorem 13. Suppose ϕ and ψ are Schwartz functions, ψ has nonvanishing integral, and $\lambda > 0$. Suppose x_0 is a point in R^n such that $N_\lambda(F_\phi, x_0) < \infty$ and so that $F_\phi(x,t)$ converges nontangentially to $A \int_{R^n} \phi$ at x_0. Then $F_\psi(x,t)$ converges nontangentially to $A \int_{R^n} \psi$ at x_0.

Theorem 14. Let $\phi \in S(R^n)$ and $\psi = \nabla\phi = (\psi_1, \ldots, \psi_n)$. Assume that for some $0 < p < \infty$ and $\lambda > 0$, x_0 is a point in R^n such that $g^*_{\lambda,p}(F_\psi, x_0) < \infty$ and $F_\phi(x,t)$ converges radially at x_0. Then, $F_\phi(x,t)$ converges nontangentially at x_0.

We remark that Theorems 8–14, with the exception of Theorem 10, hold with $F_\phi(x,t)$ and $F_\psi(x,t)$ replaced by $F_{\phi,r}(x,t)$ and $F_{\psi,r}(x,t)$ respectively, where $F_{\phi,r}(x,t) = \chi_{[0,r]}(t)F_\phi(x,t)$.

Sketch of the proof of Theorems 8–14. Since the proofs follow rather easily from the estimates obtained above, we only sketch them; Theorem 11 requires an additional argument.

Proof of Theorem 8. Since $|F_\psi(y,s)| \leq M_0(F_\psi, y)$ we can apply Theorem 1 (or 2) integrating this estimate first over s. Then, since ν is doubling, the conclusion follows at once from (the weighted version)

$$|F_\psi(x,t)|^p \leq \frac{1}{\nu(B(x,t))} \int_{R^n} M_0(F_\phi, y)^p (1 + (|x-y|/t))^{-Np} w(y)\,dy. \quad \blacksquare$$

Proof of Theorems 9 and 10. We use now the estimate in Theorem 1, 2, resp. 4, in the definition of $g^*_{\lambda_0, p_0}(F_\psi, x)$. As in the proof of Theorem 4, the desired conclusion follows using Minkowski's inequality. ∎

Proof of Theorem 11. For $\varepsilon > 0$ let

$$F^\varepsilon(x, t) = \inf_{|y - x| < \varepsilon} |F_\phi(y, t)|.$$

We need the following result concerning this function.

Lemma 15. Let ϕ be as in Theorem 2, $0 < p \le \infty$, and $\lambda > 0$. Then, there exist a number $\varepsilon_\lambda > 0$ and a constant $c_\lambda > 0$ such that if x_0 is a point in R^n for which

$$g^*_{\lambda, p}(F^\varepsilon, x_0) < \infty, \quad \text{all } \varepsilon > 0,$$

then

$$g^*_{\lambda, p}(F, x_0) \le c_\lambda g^*_{\lambda, p}(F^\varepsilon, x_0), \quad 0 < \varepsilon \le \varepsilon_\lambda.$$

The same conclusion holds with $g^*_{\lambda, p}$ replaced by N_λ and g_p replaced by M_0 above.

Assume for the moment that the lemma has been proved. Observe, then, that

$$\varepsilon^n S_{\varepsilon, p}(F^\varepsilon, x) \le g_p(F, x), \quad \text{all } \varepsilon > 0.$$

Whence, from results in Chapter IV we get that

$$\|g^*_{\lambda, p}(F^\varepsilon)\|_{L^q_w} \le c_\varepsilon \|\varepsilon^n S_{\varepsilon, p}(F^\varepsilon)\|_{L^q_w} \le c_\varepsilon \|g_p\|_{L^q_w},$$

and consequently, we conclude that, except possibly in a subset of R^n with ν measure 0, we have that $g^*_{\lambda, p}(F^\varepsilon, x) < \infty$ for all $\varepsilon > 0$. By Lemma 7 it then follows that, except on a set of ν measure 0, $g^*_{\lambda, p}(F, x) \le c_\lambda g^*_{\lambda, p}(F^\varepsilon \lambda, x)$. Thus,

$$\|g^*_{\lambda, p}(F)\|_{L^q_w} \le c_\lambda \|g^*_{\lambda, p}(F^\varepsilon \lambda)\|_{L^q_w} \le c_\lambda c_\varepsilon \|g_p(F)\|_{L^q_w},$$

which is the desired conclusion. Since the case with $g^*_{\lambda, p}$ and g_p replaced by N_λ and M_0 respectively, is proved in an analogous fashion, it only remains to prove the lemma.

Proof of Lemma 15. We recall the proof of the weighted version of Theorem 1. By assuming that

$$\tilde{M}_N(x, t) = \sup_{y \in R^n, \beta^j < a} |F_\phi(y, \beta^j t)| (1 + (|x - y|/t))^{-N} \beta^{j^N},$$

was finite, we could find a point $x_0 \in R^n$ and an integer j_0 such that

$$|F_\phi(x_0, \beta^{j_0} t)| (1 + (|x - x_0|/t))^{-N} \ge \frac{1}{4} \tilde{M}_N(x, t)$$

for all $y \in B(x_0, c_N \beta^{j_0} t)$. From this it follows that Theorem 2 holds with $F_\phi(y,s)$ replaced by $|F_\phi^\epsilon(y,s)|$ with $\epsilon \leq c_N/2$, provided that N is chosen so large that $\tilde{M}_N(x,t)$ is finite for all (x,t). As in Theorem 9, we now get that $N_\lambda(F_\phi, x_0) \leq cg_{\lambda,p}^*(F_\phi, x_0) \leq c_N g^a st_{\lambda,p}(F_\phi^\epsilon, x_0) < \infty$, $0 < \epsilon < c_N/2$, provided N is large enough, the exact choice depending on the tempered distribution f. This means that $\tilde{M}_N(x,t)$ is finite for all $(x,t) \in R_+^{n+1}$. Thus,

$$g_{\lambda,p}^*(F_\phi, x_0) \leq c_\lambda g_{\lambda,p}^*(F_\phi^\epsilon, x_0), \quad \text{all } 0 < \epsilon < \epsilon_\lambda = c_\lambda/2,$$

and c_λ does not depend on f nor does it depend on x_0. This completes the proof of the lemma. ∎

Proof of Theorem 12. If $g_{\lambda,p}^*(F_\phi, x_0) < \infty$, then

$$g_{\lambda,p}^*(F_\phi, r, x_0)^p = \int_0^r \int_{R^n} |F_\phi(y,t)|^p (1 + (|x_0 - y|/t))^{-\lambda p} t^{-n} \, dy \frac{dt}{t}$$

tends to 0 as $r \to 0$. As in Theorem 6 we have

$$N_\lambda(F_{\phi,r}, x_0) \leq cg_{\lambda,p}^*(F_{\phi,cr}, x_0),$$

and consequently, $N_\lambda(F_{\phi,r}, x_0) \to 0$ wtih r; this shows the nontangential convergence. ∎

Proof of Theorem 13. By subtracting a constant function from f we may assume that $A = 0$. Note that if $\lambda_1 > \lambda$, then $N_{\lambda_1}(F_{\psi,r}, x_0)$ is dominated by

$$cN_{\lambda_1}(F_{\phi,cr}, x_0) + c \sup_{y \in R^n, 0 < t < cr} |F_\phi(y,t)|(1 + (|x_0 - y|/t))^{\lambda_1}$$

$$\leq cN_\lambda(F_\phi, x_0) \sup_{|y-x_0| > ht} (1 + (|x_0 - y|/t))^{-\lambda_1 + \lambda}$$

$$+ \sup_{|y-x_0| \leq ht, 0 < t \leq cr} |F_\phi(y,t)|(1 + (|y - x_0|/t))^{\lambda_1}$$

$$\leq c(1 + h)^{-(\lambda_1 - \lambda)} N_\lambda(F_\phi, x_0) + \sup_{|y-x_0| < ht < chr} |F_\phi(y,t)|.$$

Thus, by first picking h large enough and then r small enough, we can make this expression arbitrarily small. Since $N_{\lambda_1}(F_{\psi,r}, x_0)$ goes to 0 with r, we have shown that $F_\psi(x,t)$ converges nontangentially to 0. This completes the proof. ∎

Proof of Theorem 14. We have $t\nabla F_\phi(x,t) = F_\psi(x,t)$. Whence, by the mean value theorem it follows that

$$|F_\phi(y,t) - F_\phi(x_0,t)| \leq c_h N_\lambda(F_{\psi,r}, x_0), \quad \text{for } |y - x_0| < ht.$$

Thus,

$$\sup_{|y-x_0|<ht<h\varepsilon} |F_\phi(y,t) - A|$$

$$\leq c_h N_\lambda(F_{\psi,\varepsilon}, x_0) + \sup_{0<t<\varepsilon} |F_\phi(x_0,t) - A|.$$

By the assumption concernaing the radial convergence the last term in the above sum goes to 0 with ε. As for the first term, since

$$N_\lambda(F_{\psi,\varepsilon}, x_0) \leq c g^*_{\lambda,p}(F_{\psi,c\varepsilon}, x_0),$$

and since $g^*_{\lambda,p}(F_\psi, x_0)$ is finite, it also goes to 0 with ε. ■

A word about the constants that appear on the right-hand side of Theorems 1 and 2 above: They can be made independent of the function ϕ there, once we normalize these class of functions is properly normalized. Specifically, let

$$\mathcal{A} = \{\phi \in \mathcal{S}(R^n): \int_{R^n} (1+|x|)^{N_0} \Big(\sum_{|\alpha|\leq N_0} \Big| \frac{\partial^\alpha}{\partial x^\alpha} \phi(x) \Big|^2 \Big) dx \leq 1\},$$

where N_0 is a large number depending on n. Then, the constants there are independent of $\phi \in \mathcal{A}$.

Sources and Remarks. The idea of representing F_ψ as a convolution of F_ϕ with a measure $d\mu(y,t) = g(y,t)dy\frac{dt}{t}$ in the upper half-space with, roughly speaking, $|g(y,t)| \leq c_N(1+(|y|/t))^{-N}t^N$ for $0 < t \leq 1$ and $\leq (1+(|x|/t))t^{-(1-1/N)}$ for $t > 1$ when the integral of ϕ vanishes, and $= 0$ if the integral of ϕ does not vanish, for every $N > 0$, is essentially due to C. Fefferman and E. Stein [1972], where the support of the measure μ is contained on the lines $\{(y,t) \in R_+^{n+1} : t = 2^{-k}, k = 0, 1, \ldots\}$, and to A. Calderón and A. Torchinsky [1975] where the support of the measure μ is contained in the upper half-space. The latter authors also introduced the condition (1) to replace the single function ϕ with nonvanishing integral. The estimates in Theorems 1–6 are extensions of the mean value properties of harmonic, subharmonic and temperature functions, where somewhat sharper results can be obtained.

CHAPTER

VI

The Hardy Spaces

Previous chapters were preliminary in the sense that they covered some of the basic results leading to the theory of the weighted Hardy spaces. Another important property, namely, the fact that the "norm" of a distribution in these spaces can be computed in various equivalent ways, is the content of this chapter.

Given a tempered distribution $f \in \mathcal{S}'(R^n)$ and a Schwartz function $\phi \in \mathcal{S}(R^n)$ with nonvanishing integral, let $F_\phi(x,t) = f * \phi_t(x)$. Assume that ν is a doubling weighted measure with respect to the Lebesgue measure in R^n with weight w, let $a > 0$, and for $0 < p < \infty$ set

$$H_w^p(R^n) = \{f \in \mathcal{S}'(R^n) : M_a(F_\phi) \in L_w^p(R^n)\}, \quad \|f\|_{H_w^p} = \|M_a(F_\phi)\|_{L_w^p}.$$

We leave it up to the reader to show the basic topological properties of these spaces; for instance, with the distance function $d(f,g) = \|f - g\|_{H_w^p}$, $H_w^p(R^n)$ becomes a complete metric space when $0 < p \le 1$, etc.

Now, according to Theorem 10 in Chapter IV, $\|M_b(F_\phi)\|_{L_w^p}$ is equivalent to $\|f\|_{H_w^p}$ for all $b > 0$, and also for $b = 0$ under the assumptions of Theorem 12 in that chapter. Furthermore, Theorem 6 in Chapter V readily implies that the same is true for $\|M_b(F_\psi)\|_{L_w^p}$ provided that ψ is a Schwartz function with nonvanishing integral. Indeed, let $0 < p_0 < p$ and note that

$$\|M_1(F_\psi)\|_{L_w^p} = \|M_1(F_\psi)^{p_0}\|_{L_w^{p/p_0}}^{1/p_0} \le c\|M_\nu(M_0(F_\phi)^{p_0})\|_{L_w^{p/p_0}}^{1/p_0}.$$

Since ν is doubling and $p/p_0 > 1$, by Theorem 3 in Chapter I, we see that the above expression is dominated by

$$c\|M_0(F_\phi)^{p_0}\|_{L_w^{p/p_0}}^{1/p_0} = c\|M_0(F_\phi)\|_{L_w^p},$$

and consequently, $M_1(F_\psi) \in L^p_w(R^n)$. By exchanging the roles of ϕ and ψ we see that the Hardy spaces are intrinsically defined, i.e., they are independent of the approximate identity ϕ and the opening $a > 0$ of the cone used in the definition of $M_a(F_\phi, x)$.

In fact, even more is true. Referring to the comment closing Chapter V, given $f \in \mathcal{S}'(R^n)$, consider the "grand maximal function"

$$f^*(x) = \sup_{\phi \in \mathcal{A}} \sup_{|x-y|<t} |f * \phi_t(y)| \, .$$

Then, the above comments imply that if N_0 in the definition of $\mathcal{A}$ is a large number depending only on n, p, and w, $f^* \in L^p_w(R^n)$ if and only if $M_1(F_\phi) \in L^p_w(R^n)$, where ϕ is some function in $\mathcal{S}(R^n)$ with nonvanishing integral. Thus, there is also a characterization of $H^p_w(R^n)$ in terms of grand maximal functions.

We point out that, even when $p > 1$, the weighted Hardy spaces may contain measures and distributions. Indeed, given $1 \leq p < q$, it is possible to construct a weight $w \in A_q(R)$ such that the Dirac δ measure concentrated at the origin is in $H^p_w(R)$. Let $w(x) = |x|^a/(1 + |x|)^b$, then $w \in A_r$ for $-1 < a < r - 1$ and $-1 < a - b < r - 1$. Since for a Schwartz function ϕ we have $M_1(\delta * \phi_t, x) \leq c/|x|$, it follows that

$$\int_R M_1(\delta * \phi_t, x)^p w(x)\, dx < \infty \, ,$$

provided that $a > p - 1$ and $p + b - a > 1$. By first picking $q - 1 > a > p - 1$ and then b so that $p + b - a > 1$ and $-1 < a - b$, all conditions are met.

The situation is altogether different when $w \in A_p$.

Theorem 1. Suppose f is a tempered distribution in $H^p_w(R^n)$, $w \in A_p$, for $1 < p < \infty$. Then there is a function $g \in L^p_w(R^n)$ such that $F_\phi(x,t) = g * \phi_t(x)$ for all Schwartz functions ϕ, and $\|f\|_{H^p_w} \sim \|g\|_{L^p_w}$.

In applications it is often useful to consider the Lusin and Littlewood-Paley functions associated with $f \in H^p_w(R^n)$; it is, therefore, an important result in the theory that an equivalent norm of f is obtained with these functions. In order to prove this fact without invoking the atomic decomposition, we make use of the extension $F(x,t) = f * \phi_t(x)$, where $\phi(x) = e^{-\pi|x|^2}$. Note that these extensions satisfy the "heat" equation

$$\frac{\partial}{\partial t} F(x,t) - \frac{1}{2\pi} t \Delta F(x,t) = 0 \, , \quad x \in R^n, \, t > 0 \, ,$$

and are, consequently, known to satisfy a stronger version of the mean value inequality discussed in Chapter V, to wit,

$$|F(x,t)|^2 \leq c_\epsilon t^{-n} \int_{t/2}^t \int_{|x-t|\leq\epsilon t} |F(y,s)|^2\, dy \frac{ds}{s} \, .$$

Moreover, if $G(x,t) = t|\nabla F(x,t)|$, then the above estimate holds with F replaced by G either on the left-hand side or on both sides; similarly for derivatives of higher order.

If in addition $w \in A_\infty$, then the following results hold.

Theorem 2. Assume $w \in A_\infty$, and let f be a tempered distribution in $H_w^p(R^n)$, $0 < p < \infty$. If $\phi_t(x) = t^{-n}e^{-\pi(|x|/t)^2}$, $F(x,t) = f * \phi_t(x)$ and $G(x,t) = t|\nabla F(x,t)|$, for $0 < a < b < \infty$, we have

$$\|S_a(G)\|_{L_w^p} \leq c\|M_b(F)\|_{L_w^p},$$

where c depends on a,b,n,p and w, but is independent of f.

Conversely, and with the notation of Theorem 2, we have

Theorem 3. Assume $w \in A_\infty$, and let f be a tempered distribution such that $F(x,t) \to 0$ as $t \to \infty$. Then for $0 < a < b < \infty$ we have

$$\|M_a(F)\|_{L_w^p} \leq c\|S_b(G)\|_{L_w^p}, \quad 0 < p < \infty,$$

where c depends on a,b,n,p and w, but is independent of f.

A characterization of the weighted Hardy spaces in terms of singular integral operators along the lines discussed by R. Coifman and B. Dahlberg [1979] can be stated as follows. Let Γ_i, $1 \leq i \leq k$, be a collection of cones in R^n, each with vertex at the origin, given by

$$\Gamma_i = \{\xi : \xi = t\xi', t > 0, |\xi'| = 1, |\xi' - \xi_i'| < r_i\},$$

where $|\xi_i'| = 1$ and $0 < r_i < \sqrt{2}$, with the property that $R^n = \bigcup_i \Gamma_i \cup \{0\}$. Furthermore, let m_i, $1 \leq i \leq k$, be homogenous functions of degree 0, which are $C_0^\infty(R^n \setminus \{0\})$, and so that $\sum_{i=1}^k m_i(\xi) = 1$ for $\xi \neq 0$.

We then characterize the Hardy spaces by the multiplier operators T_i, $1 \leq i \leq k$, cf. Chapter XI, defined by

$$(T_if)\hat{}(\xi) = m_i(\xi)\hat{f}(\xi), \quad f \in \hat{\mathcal{D}}_0.$$

Theorem 4. Assume $w \in A_\infty$, and let $f \in \hat{\mathcal{D}}_0$. Then there is a constant c independent of f such that

$$\frac{1}{c}\sum_{i=1}^k \|T_if\|_{L_w^p} \leq \|f\|_{H_w^p} \leq c\sum_{i=1}^k \|T_if\|_{L_w^p}.$$

Theorem 5. Suppose $w \in A_\infty$. A tempered distribution f is in $H^p_w(R^n)$ if and only if there are tempered distributions f_i, $1 \leq i \leq k$, such that $\hat{f}_i(\xi) = m_i(\xi)\hat{f}(\xi)$, $\xi \neq 0$, in the sense of distributions, i.e., $\langle f_i, g \rangle = \langle f, T_i g \rangle$ for all $g \in \hat{\mathcal{D}}_0$, $f = \sum_{i=1}^k f_i$, with the property that

$$L_i = \liminf_{\varepsilon \to 0} \|f_i * \phi_\varepsilon\|_{L^p_w} < \infty, \quad i = 1, \ldots, k.$$

Here $\phi_\varepsilon(x) = \varepsilon^{-n}\phi(x/\varepsilon)$ are the dilates of a Schwartz function ϕ with nonvanishing integral. Furthermore, there is a constant c such that

$$\frac{1}{c}\sum_{i=1}^k L_i \leq \|f\|_{H^p_w} \leq c\sum_{i=1}^k L_i.$$

Theorems 4 and 5 hold for "diagonal" nonisotropic dilations under the following conditions: If $\delta_t(x_1, \ldots, x_n) = (t^{\gamma_1}x_1, \ldots t^{\gamma_n}x_n)$, then the cones are centered along the positive or negative coordinate axis. Specifically, they are of the form

$$\Gamma_{i\pm} = \{\xi : \xi = \delta_t(\xi'), t > 0, |\xi'| = 1, \xi' = (\xi'_1, \ldots, \xi'_n), \pm\xi'_i > c > 0\}.$$

We pass now to the proofs.

Proof of Theorem 1. As we have seen in Chapter V, $F_\phi(y,t)$ tends to $g(x)$ ν-a.e. as (y,t) tends to x nontangentially. Moreover, since $|g(x)| \leq M_1(F_\phi, x)$, we have $\|g\|_{L^p_w} \leq c\|f\|_{H^p_w}$, and by the Lebesgue dominated convergence theorem, $\lim_{t \to 0} \|F_\phi(\cdot, t) - g\|_{L^p_w} = 0$.

Now, since $w \in A_p$ implies $w^{-1/(p-1)} \in A_{p'}$, $1/p + 1/p' = 1$, for any Schwartz function ψ and $t > 0$ we have $\psi_t \in L^{p'}_{w^{-1/(p-1)}}(R^n) \sim L^p_w(R^n)^*$. Thus, since $F_\phi(\cdot, s) \to g$ in $L^p_w(R^n)$, it follows that $F_\phi(\cdot, s) * \psi_t(y) \to g * \psi_t(y)$ as $s \to 0$. Also, since $f * \psi_t \in C^\infty(R^n) \cap L^p_w(R^n)$, $F_\phi(\cdot, s) * \psi_t(y) = (f * \psi_t) * \phi_s(y) \to f * \psi_t(y)$ as $s \to 0$. Whence, $g * \psi_t = f * \psi_t$ for all Schwartz functions ψ.

As for the norm, we have already seen that $\|g\|_{L^p_w} \leq c\|f\|_{H^p_w}$. Moreover, since $w \in A_p$, it follows that $\|M_1(f * \psi_t)\|_{L^p_w} = \|M_1(g * \psi_t)\|_{L^p_w} \leq c\|g\|_{L^p_w}$. $\blacksquare$

Now we introduce some notations. For $0 \leq a$, $0 \leq \varepsilon < T \leq \infty$, we set

$$\Gamma^{\varepsilon,T}_a(x) = \{(y,t) : |x - y| < at, \varepsilon < t < T\}, \quad \Gamma^{0,\infty}_a(x) = \Gamma_a(x).$$

Associated to these "truncated" and full cones in R^{n+1}_+ we introduce

$$M^{\varepsilon,T}_a(F,x) = \sup_{(y,t) \in \Gamma^{\varepsilon,T}_a(x)} |F(y,t)|,$$

and with v as usual the volume of the unit ball in R^n,

$$S_a^{\varepsilon,T}(G,x) = \left(\frac{1}{v} \int \int_{\Gamma_a^{\varepsilon,T}(x)} |G(y,t)|^2 (at)^{-n} dy \frac{dt}{t} \right)^{1/2}.$$

When $\varepsilon = 0$ or $T = \infty$ we omit the corresponding superscript in the above notations; a similar remark applies to other definitions introduced in this chapter.

Proof of Theorem 3. Let

$$\tilde{M}_a^{\varepsilon,T}(F,x) = \sup_{(y_1,t_1),(y_2,t_2)\in\Gamma_a^{\varepsilon,T}(x)} |F(y_1,t_1) - F(y_2,t_2)|$$

denote the maximal oscillation of the truncations of F. Clearly $M_a(F,x) \leq \tilde{M}_a(F,x) \leq 2M_a(F,x)$ whenever $F(x,t) \to 0$ as $t \to \infty$; this is the case under our assumptions. Since $\chi_{B(0,T)}(x)\tilde{M}_a^{\varepsilon,T}(F,x)$ increases to $\tilde{M}_a(F,x)$ for each $x \in R^n$ as $\varepsilon \to 0$ and $T \to \infty$, the desired conclusion follows from the estimate

$$\|\chi_{B(0,T)}\tilde{M}_a^{\varepsilon,T}(F)\|_{L_w^p} \leq c\|S_b(G)\|_{L_w^p}, \tag{1}$$

with c independent of ε, T and f.

We first prove a localization result that lies at the heart of the matter.

Lemma 6. Let $0 < \eta < 1$ be given. For $\delta, s > 0$ and $0 < \varepsilon < T < \infty, 0 < a < b,$ $\lambda > 0$, and x in R^n, put

$$\mathcal{E} = \{z \in B(x,T): \tilde{M}_a^{\varepsilon,T}(F,z) > \lambda, M_{0,s}(S_b^{\varepsilon/2,2T}(G))(z) < \delta\lambda\}.$$

Then we can choose δ, s independent of ε, T, λ and $B(x,T)$ so that

$$|\mathcal{E}| \leq \eta|B(x,T)|.$$

Proof. Since subtracting a constant from F leaves both $\tilde{M}_a^{\varepsilon,T}(F)$ and $S_b^{\varepsilon/2,2T}(G)$ unchanged, we may assume that $F(x,T) = 0$. If $\mathcal{E} = \emptyset$ there is nothing to prove. Otherwise, since for $z \in \mathcal{E}$ we have $M_a^{\varepsilon,T}(F,z) > \lambda/2$, to each $z \in \mathcal{E}$ we can assign $(y_z,t_z) \in \Gamma_a^{\varepsilon,T}(z)$ such that $|F(y_z,t_z)| > \lambda/2$. Since $t_z \leq T$ we can find a sequence $\{B(y_k,at_k)\}$ of pairwise disjoint balls corresponding to points $z_k \in \mathcal{E}$ so that

$$\mathcal{E} \subseteq \bigcup_k B^1(y_k,at_k). \tag{2}$$

Moreover, since for each k we have $B(y_k,at_k) \subseteq B(x,(a+1)T)$ and $t_k > \varepsilon > 0$, there can only be a finite number N, say, of such balls. Also, since

$\{S_b^{\epsilon/2,2T}(G) \le \mu\} \supseteq \{M_{0,s}(S_b^{\epsilon/2,2T}(G)) \le \mu\}$ for each $\mu > 0$, for $z \in \mathcal{E}$ we have $S_b^{\epsilon/2,2T}(G, z) \le \delta\lambda$.

Given $0 < t < T$, let $\Omega(y, t)$ denote the cylinder $\{(y', s) \in R_+^{n+1} : |y - y'| < (a + b)t/2, t/2 \le s \le t\}$. By the mean value inequality for the derivatives of solutions to the heat equation we have

$$G(y, t)^2 \le c \iint_{\Omega(y,t)} G(y', s)^2 s^{-n} dy' \frac{ds}{s}. \tag{3}$$

A moment's thought will convince the reader that if $z \in \mathcal{E}$ and $|y - z| < at$, then $\Omega(y, t) \subseteq \Gamma_b^{\epsilon/2,2T}(z)$. Therefore from (3) it follows that

$$G(y, t) \le cb^n \iint_{\Gamma_b^{\epsilon/2,2T}(z)} G(y', s)^2 (bs)^{-n} dy' \frac{ds}{s}$$

$$\le cb^n S_b^{\epsilon/2,2T}(G, s)^2 \le c(\delta\lambda)^2, \tag{4}$$

where the constant c depends on b, n, but it is independent of f.

Let now $T_a(y, t)$ denote the tent, or inverted cone, $\{(y', s) \in R_+^{n+1} : |y' - y| < a(t - s)\}$, $\partial T_a(y, t)$ denote its boundary, and σ denote the surface area in R_+^{n+1}. We claim we can choose $\delta > 0$ sufficiently small so that

$$|B(y_k, at_k)| \le c\sigma(\{(y', s) \in \partial T_a(y_k, t_k) : |F(y', s)| > \lambda/3\}), \quad \text{all } k. \tag{5}$$

Indeed, let $y \in B(z, aT)$, $z \in \mathcal{E}$. By the mean value inequality, with the left-hand side replaced by higher order derivatives of F such as $s^2 \Delta F(y, s)$, and (3), it follows that for $t_1/t_2 \sim 1$,

$$|F(y, t_1) - F(y, t_2)| \le \int_{t_2}^{t_1} \left| \frac{\partial}{\partial s} F(y, s) \right| ds$$

$$= \int_{t_2}^{t_1} |s^2 \Delta F(y, s)| \frac{ds}{s} \le c(\delta\lambda) \ln(t_1/t_2) \le c(\delta\lambda). \tag{6}$$

Also, if $(y, t), (y', t)$ are in $\Gamma_b^{\epsilon/2,2T}(z)$ and $z \in \mathcal{E}$, we have

$$|F(y, t) - F(y', t)| = \left| \int_0^1 \left\langle t\nabla F(y' + s(y - y'), t), \frac{y - y'}{t} \right\rangle ds \right|$$

$$\le \frac{|y - y'|}{t} \int_0^1 G(y' + s(y - y'), t) \, ds \le c(|y - y'|/t)(\delta\lambda). \tag{7}$$

We combine these estimates. Since $|F(y_k, t_k)| > \lambda/2$, from (6) it follows that $|F(y_k, t)| > \lambda/2 - c\delta\lambda$ for $t_k/t \sim 1$ and, by (7), $|F(y, t_k)| > \lambda/2 - c'\delta\lambda$ whenever

$(y, t_k) \in \Gamma_b^{\varepsilon/2, 2T}(z_k)$. Therefore, $|F| > \lambda/3$ in a fixed proportion of $\partial T_a(y_k, t_k)$, independently of k. Whence, since $|B(y_k, at_k)| \sim \sigma(\partial T_a(y_k, t_k))$, (5) holds.

We are now ready to estimate $|\mathcal{E}|$. Let

$$\mathcal{U} = \left(\bigcup_{k=1}^{N} \Gamma_{(a+b)/2}^{2\varepsilon/3, 3T/2}(z_k) \right) \setminus \left(\bigcup_{k=1}^{N} T_a(y_k, t_k) \right).$$

The boundary $\partial \mathcal{U}$ of $\mathcal{U}$ consists of two parts, namely

$$\partial_1 \mathcal{U} = \{(y, s) \in \partial \mathcal{U} : 2\varepsilon/3 \leq s \leq 3T/2\},$$

and

$$\partial \mathcal{U}_2 = \{(y, s) \in \partial \mathcal{U} : s = 3T/2\}.$$

By (2) and (5) we conclude that

$$|\mathcal{E}| \leq c\lambda^{-2} \int_{\partial_1 \mathcal{U}} |F(y, t)|^2 \, d\sigma(y, t). \tag{8}$$

Next we estimate the right-hand side of (8). With no loss of generality we may assume that F is real valued. We begin by considering the integral $\int \int_{\mathcal{U}} G(y, t)^2 \, dy \frac{dt}{t}$. Observe that if $(y, t) \in \Gamma_{3a/2}^{2\varepsilon/3, 3T/2}(z_k)$, there is a ball $B(z_k, cat)$, with c sufficiently small, so that $(y, t) \in \Gamma_b^{\varepsilon/2, 2T}(y')$ for every $y' \in B(z_k, cat)$. Since $M_{0,s}(S_b^{\varepsilon/2, 2T}(G))(z_k) \leq \delta\lambda$, it follows that

$$|\{y' \in B(z_k, cat) : S_b^{\varepsilon/2, 2T}(G, y') \leq \delta\lambda\}| \geq (1 - s)|B(z_k, cat)|$$
$$= (1 - s)c(at)^n.$$

Thus, if $\mathcal{E}_1 = \{y \in B(x, T) : S_b^{\varepsilon/2, 2T}(G, y) < \delta\lambda\}$, we have

$$\int \int_{\mathcal{U}} G(y, t)^2 \, dy \frac{dt}{t} \leq c \int_{\mathcal{E}_1} S_b^{\varepsilon/2, 2T}(G, y)^2 \, dy$$
$$\leq c(\delta\lambda)^2 |B(y, T)|. \tag{9}$$

On the other hand, from (4) it follows that

$$G(y, t) \leq c\delta\lambda, \quad \text{whenever } (y, t) \in \mathcal{U}. \tag{10}$$

Moreover, since $F(x, T) = 0$, again from the mean value inequality, we conclude that

$$|F(y, t)$$

Let $n(y,t)$ denote the normal unit vector to $\mathcal{U}$ at the point (y,t) of the boundary of $\mathcal{U}$; $n(y,t)$ is defined σ-a.e. Clearly, $n(y,t) = (0,\ldots,0,-1)$ on $\partial\mathcal{U}_2$, and $n(y,t) = (n_1(y,t),\ldots,n_{n+1}(y,t))$ with $n_{n+1}(y,t) \geq ca^{-1}$ on $\partial\mathcal{U}_1$. Thus,

$$d\sigma(y,t) \leq c_a dy, \quad \text{a.e. } (y,t) \in \partial\mathcal{U}. \tag{12}$$

Let $\mathcal{U}_t = \mathcal{U} \cap \{(y,t) \in R_+^{n+1} : t \text{ fixed}\}$, and let $d\tau_t(y,t)$ denote the surface measure on the $(n{-}1)$-dimensional boundary $\partial\mathcal{U}_t$ of $\mathcal{U}_t$. Then,

$$d\tau_t(y,t)dt \leq d\sigma(y,t) \text{ a.e. } (y,t) \in \partial_1\mathcal{U}.$$

If $\tilde{\mathcal{U}} = \{y \in R^n : (y,t) \in \mathcal{U}\}$, to each $y \in \tilde{\mathcal{U}}$ there corresponds exactly one point $(y,t_1(y)) \in \partial_1\mathcal{U}$ and one point $(y,t_2(y)) \in \partial_2\mathcal{U}$.

We are ready to estimate $\int_{\partial_1\mathcal{U}} F(y,t)^2\, d\sigma(y,t)$. For any y in $\tilde{\mathcal{U}}$,

$$F(y,t_2(y))^2 - F(y,t_1(y))^2 = \int_{t_1(y)}^{t_2(y)} \frac{\partial}{\partial t} F(y,t)^2\, dt.$$

Whence integrating over $\tilde{\mathcal{U}}$ with respect to dy by (12) we get

$$\int_{\partial_2\mathcal{U}} F(y,t)^2\, d\sigma(y,t) - c_a^{-1} \int_{\partial_1\mathcal{U}} F(y,t)^2\, d\sigma(y,t) \geq \int_{\mathcal{U}} \frac{\partial}{\partial t} F(y,t)^2\, dy dt.$$

Since F satisfies the heat equation we have

$$\frac{\partial}{\partial t} F(y,t)^2 = 2F(y,t)\frac{\partial}{\partial t}F(y,t) = \frac{1}{\pi} F(y,t)t\Delta F(y,t).$$

Substituing this relation in the right-hand side above, by (11) it follows that

$$\int_{\partial_1\mathcal{U}} F(y,t)^2\, d\sigma(y,t) \leq c_a(\delta\lambda)^2\sigma(\partial_2\mathcal{U})$$

$$+ c\int\int_{\mathcal{U}} tF(y,t)\Delta F(y,t)\, dy dt = I + J, \tag{13}$$

say. To estimate J we fix t and consider the n-dimensional set $\mathcal{U}_t$. $\mathcal{U}_t$ is the union of finitely many balls minus the union of finitely many pairwise disjoint balls, and, as such, it can be approximated by smooth regions where we can apply Stokes' theorem to the identity

$$\nabla\cdot(F\nabla F) = |\nabla F|^2 + F\Delta F.$$

Hence

$$\left| \int_{\mathcal{U}_t} F(y,t)\Delta F(y,t)\, dy \right| \leq \int_{\mathcal{U}_t} |\nabla F(y,t)|^2\, dy$$

$$+ \int_{\partial\mathcal{U}_t} |F(y,t)|\,|\nabla F(y,t)|\, d\tau_t(y,t). \tag{14}$$

By (10) and the inequality $2AB \leq \delta A^2 + \delta^{-1}B^2$ applied to the second integral above, upon multiplying by t and integrating we get

$$|J| \leq \int \int_{\mathcal{U}} G(y,t)^2 \, dy \frac{dt}{t} + c\delta \lambda^2 \sigma(\partial_1 \mathcal{U}) + \delta \int_{\partial_1 \mathcal{U}} F(y,t)^2 \, d\sigma(y,t).$$

Since $\sigma(\partial_1 \mathcal{U}), \sigma(\partial_2 \mathcal{U}) \leq c|B(x,T)|$, by combining (9), (13) and (14) we get

$$\int_{\partial_1 \mathcal{U}} F(y,t)^2 \, d\sigma(y,t) \leq c(\delta + \delta^2)|B(x,T)| + \delta \int_{\partial_1 \mathcal{U}} F(y,T)^2 \, d\sigma(y,t).$$

Since $\partial_1 \mathcal{U}$ is contained in a compact set where F is continuous, the last integral above is finite. So, if $\delta < 1/2$, we can move this integral to the left-hand side of the inequality and obtain

$$\int_{\partial_1 \mathcal{U}} F(y,t)^2 \, d\sigma(y,t) \leq 2c(\delta + \delta^2)\lambda^2 |B(x,T)|.$$

From (8) we now get that $|\mathcal{E}|$ is also dominated by the right-hand side above, and the desired conclusion follows upon choosing δ sufficiently small. ∎

To prove (1) we also make use of the following good-lambda inequality.

Lemma 7. Let $0 < \eta < 1$, and for $\delta, s > 0$, $0 < \varepsilon < T < \infty$, and $a_1 < a_2$, $\lambda > 0$, let

$$\mathcal{E} = \{z \in B(0,T) : \tilde{M}_a^{\varepsilon,T}(F,z) > \lambda, M_{0,s}(S_b^{\varepsilon/2,2T}(G))(z) < \delta\lambda\}.$$

Then we can choose δ sufficiently small, independently of the other parameters, so that

$$\nu(\mathcal{E}) \leq \eta \nu(\{z \in B(0,T) : \tilde{M}_a^{\varepsilon,T}(F,z) > \lambda/2\}). \tag{15}$$

Proof. Let c_2 be the constant in Lemma 1 in Chapter I. To each $z \in \mathcal{E}$ we assign a positive number h_z with the following three properties: (i) $c_2 h_z < T$, (ii) $\tilde{M}_a^{\varepsilon,T}(F,y) > \lambda/2$ for $y \in B(z,h_z)$, and, (iii) Either $c_2 h_z > T/2$, or $\tilde{M}_a^{\varepsilon,T}(F,y_z) \leq \lambda/2$ for some $y_z \in B(z, c_2 h_z)$.

Now, for $z \in \mathcal{E}$ we also have $G(y,t) < c\delta\lambda$ for $(y,t) \in \Gamma_a^{\varepsilon,T}(z)$. Thus, if $c_2 h_z > T/c_0$, where $c_0 > 2$ is a constant to be chosen shortly, from the mean value theorem we get

$$\tilde{M}_a^{\varepsilon,T}(F,z) \leq \tilde{M}_a^{\varepsilon,c_2 h_z}(F,z) + c\delta\lambda.$$

On the other hand, if $c_2 h_z \leq T/c_0$, then there is a point $y_z \in B(z, c_2 h_z)$ such that $\tilde{M}_a^{\varepsilon,T}(F,y_z) \leq \lambda/2$, and we can pick $c_0 > 2$ so large that for each t with

$c_0 c_2 h_z < t < T$, there is a point (y,t) contained in both $\Gamma_a^{\varepsilon,T}(z)$ and $\Gamma_a^{\varepsilon,T}(y_z)$. By the mean value theorem we conclude that in this case

$$\tilde{M}_a^{\varepsilon,T}(F,z) \leq \tilde{M}_a^{\varepsilon,c_2 h_z}(F,z) + c\delta\lambda + \lambda/2\,.$$

Thus, provided that δ is sufficiently small, it follows that $\tilde{M}_a^{\varepsilon,c_2 h_z}(F,z) > \lambda/3$ for $z \in \mathcal{E}$. By a covering argument we select a pairwise disjoint subcollection $\{B(z_k, h_{z_k})\}$ of the $B(z,h_z)$'s, $z \in \mathcal{E}$, so that $\mathcal{E} \subseteq \bigcup_k B(z_k, c_2 h_{z_k})$. Whence, $\nu(\mathcal{E}) \leq \sum_k \nu(\mathcal{E}_k)$, where each for each k

$$\mathcal{E}_k = \{y \in B(z_k, c_2 h_{z_k}) \colon \tilde{M}_a^{\varepsilon,c_2 h_{z_k}}(F,y) > \lambda/3, M_{0,s}(S_b^{\varepsilon/2,2T}(G))(y) < \delta\lambda\})\,, \tag{16}$$

and

$$\sum_k \nu(B(z_k, h_{z_k})) \leq \nu(\{y \in B(0,T) \colon \tilde{M}_a^{\varepsilon,T}(F,y) > \lambda/2\})\,. \tag{17}$$

Since $w \in A_\infty$ there are a number $q > 1$ and a constant c such that

$$\frac{1}{c}(|\mathcal{U}|/|B|)^q \leq \nu(\mathcal{U})/\nu(B) \leq c(|\mathcal{U}|/|B|)^{1/q}\,,$$

for each ball B and each measurable subset $\mathcal{U}$ of B. Therefore, from Lemma 6 it follows that for some constant b,

$$\nu(\mathcal{E}_k) \leq c\eta_1^b \nu(B(z_k, c_2 h_{z_k})) \leq c\eta_1^b \nu(B(z_k, h_{z_k}))\,, \tag{18}$$

for $\eta_1 < 1$ provided δ is small enough. From (16), (18) and (17) it follows that

$$\nu(\mathcal{E}) \leq c\eta_1^b \nu(\{y \in B(0,T) \colon \tilde{M}_a^{\varepsilon,T}(F,y) > \lambda/2\})\,,$$

which is the desired estimate once we pick η_1 small enough; $c\eta_1^b < \eta$ will do. ∎

The proof of Theorem 3 now follows at once. By integrating (15) with respect to $p\lambda^{p-1}$, $0 < p < \infty$, we obtain

$$I = \|\chi_{(0,T)} \tilde{M}_a^{\varepsilon,T}(F)\|_{L_w^p}^p \leq c_{\delta,p} \|M_{0,s}(S_b^{\varepsilon/2,2T}(G))\|_{L_w^p}^p + 2^p\eta I\,.$$

Since $\chi_{B(0,T)} \tilde{M}_a^{0,T}(F)$ is a bounded function of compact support and ν is finite on bounded sets, $I < \infty$. Thus, if η in Lemma 7 is small enough, $\eta < 2^{-p-1}$ will do, by Lemma 1 in Chapter III we get

$$I \leq c_{\delta,p} \|M_{0,s}(S_b^{\varepsilon/2,2T}(G))\|_{L_w^p}^p \leq c\|S_b^{\varepsilon/2,2T}(G)\|_{L_w^p}^p\,,$$

and we are done. ∎

As for the proof of Theorem 2, we begin by discussing some preliminary results.

Lemma 8. Assume $\mathcal{O} \subset R^n$ is a proper open subset of R^n, and let $\tilde{d}(x)$ denote the distance of $x \in R^n$ to the complement of $\mathcal{O}$. If $\theta \in C_0^\infty(R^n)$ is supported in $B(0, 1/2)$ and has integral 1, consider the functions

$$\Omega(x,t) = \begin{cases} 1 & \text{if } t \geq \tilde{d}(x) \\ 0 & \text{otherwise,} \end{cases}$$

$$\Omega_1(x,t) = \begin{cases} 1 & \text{if } t \geq \tilde{d}(x) \\ 2t\tilde{d}(x)^{-1} - 1 & \text{if } \tilde{d}(x)/2 \leq t \leq \tilde{d}(x) \\ 0 & \text{otherwise,} \end{cases}$$

and

$$\Omega_2(x,t) = \int_{R^n} \Omega(y, 3t/2)\theta_t(x-y)\, dy\,.$$

Then we have

(i) $\int_0^\infty \frac{\partial}{\partial t}\Omega_1(x,t)\, dt = 1$ if $x \in \mathcal{O}$, and 0 otherwise.

(ii) $\Omega(x,2t) - \Omega(x,t) \leq t\frac{\partial}{\partial t}\Omega_1(x,t) \leq 2(\Omega(x,2t) - \Omega(x,t))$.

(iii) $|\Omega_1(x,t) - \Omega_2(x,t)| \leq 2t\frac{\partial}{\partial t}\Omega_1(x,t)$, and similarly for $|\Omega - \Omega_1|$ and $|\Omega - \Omega_2|$.

(iv) $|D^\sigma \Omega_2(x,t)| \leq c_\sigma t\frac{\partial}{\partial t}\Omega_1(x,t)$.

Since the proof of these results is immediate, it is left for the reader to verify.

Lemma 9. Let f be a real tempered distribution, $\phi_t(x) = t^{-n}e^{-\pi(|x|/t)^2}$, and put $\psi(x) = \Delta\phi(x)$. Suppose $\mathcal{O}$ is a nonempty subset of R^n, and let $\mathcal{E}$ denote its complement. Let Ω, Ω_1 and Ω_2 be related to $\mathcal{O}$ as in Lemma 8, and assume that either $\mathcal{E}$ is bounded or that $\hat{f}(\xi)(1 + |\xi|^2)^{-N}$ is square integrable for sufficiently large N. Let $t_1 < t_2$ be two positive numbers, and for $a > 0$ set

$$h(x)^2 = \int_{t_1}^{t_2} F(x,t)^2 \frac{\partial}{\partial t}\Omega_1(x,at)\, dt + F(x,t_1)^2\Omega_1(x,at_1)\,.$$

With the notation of Theorem 2 the following estimate holds:

$$\int_{t_1}^{t_2} \int_{R^n} G(x,t)^2 \Omega(x,at)\, dx\, \frac{dt}{t} + \int_{R^n} F(x,t_2)^2\Omega_1(x,at_2)\, dx$$

$$\leq \frac{3}{2}\int_{R^n} h(x)^2\, dx + \int_{R^n}\int_{t_1}^{t_2}(|F_\psi(x,t)|^2 + cG(x,t)^2)\frac{\partial}{\partial t}\Omega_1(x,at)\, dx\, dt\,.$$

Proof. Integrating $(\frac{\partial}{\partial t}F(x,t)^2)\Omega_1(x,at)$ between t_1 and t_2 we obtain

$$\int_{R^n} h(x)^2\, dx = \int_{R^n} F(x,t_2)^2\Omega_1(x,at_2)\, dx$$

$$- 2\int_{R^n}\int_{t_1}^{t_2} F(x,t)\frac{\partial}{\partial t}\Omega_1(x,at)\, dx\, dt = I - 2J\,, \qquad (19)$$

say. If $\mathcal{E}$ is bounded, the domain of integration of J is bounded. If $\mathcal{E}$ is not bounded, since $\hat{f}(\xi)(1 + |\xi|^2)^{-N}$ is square integrable, then $F(x,t)$ and all its derivatives, and in particular $G(x,t)$ and $F_\psi(x,t)$, are square integrable with respect to x, uniformly on every closed interval of values of t contained in $t > 0$. Thus, in all cases, J is absolutely convergent, and all integration by parts below are justified. Using the heat equation we see that

$$
\begin{aligned}
J &= \int_{t_1}^{t_2} \int_{R^n} F(x,t) t \Delta F(x,t) \Omega_1(x,at)\, dx \frac{dt}{t} \\
&= -\int_{t_1}^{t_2} \int_{R^n} F(x,t) F_\psi(x,t)(\Omega_2(x,at) - \Omega_1(x,at))\, dx \frac{dt}{t} \\
&\quad + \int_{t_1}^{t_2} \int_{R^n} F(x,t) F_\psi(x,t) \Omega_2(x,at)\, dx \frac{dt}{t}\,.
\end{aligned}
$$

Upon integrating by parts the last integral above becomes

$$
\begin{aligned}
&-\int_{t_1}^{t_2} \int_{R^n} G(x,t)^2 \Omega(x,at)\, dx \frac{dt}{t} \\
&\quad -\int_{t_1}^{t_2} \int_{R^n} G(x,t)^2 (\Omega_2(x,at) - \Omega(x,at))\, dx \frac{dt}{t} \\
&\quad -\int_{t_1}^{t_2} \int_{R^n} F(x,t) \Big(\sum_{j=1}^{n} F_{\frac{\partial \phi}{\partial x_j}}(x,t) \frac{\partial}{\partial x_j} \Omega_2(x,t) \Big)\, dx \frac{dt}{t}\,.
\end{aligned}
$$

Since $2|FF_\psi| \leq F^2/\beta + \beta F_\psi^2$ for all $\beta > 0$, by (iii) and (iv) of Lemma 8 we find that

$$
\begin{aligned}
&2\left| J + \int_{t_1}^{t_2} \int_{R^n} G(x,t)^2 \Omega(x,at)\, dx \frac{dt}{t} \right| \\
&\qquad\qquad \leq 1/2 \int_{R^n} \int_{t_1}^{t_2} F(x,t)^2 \frac{\partial}{\partial t}\Omega_1(x,at)\, dx dt \\
&\qquad\qquad\quad + \int_{R^n} \int_{t_1}^{t_2} (4F_\psi(x,t)^2 + cG(x,t)^2) \frac{\partial}{\partial t}\Omega_1(x,at)\, dx dt\,,
\end{aligned}
$$

which combined with (19) gives the desired estimate. ∎

Lemma 10. Assume the assumptions of Lemma 9 hold, and that f is merely a tempered distribution. Let $N(x) = M_{2a}(\max(|F|, G, |F_\psi|), x)$, and for a fixed compact subset K of R_+^{n+1}, set $S_{a,K}(G,x) = S_a(\chi_K G, x)$. Let $0 < \eta < 1$ be given, and for a ball B and $a > 0$, and $\lambda, \varepsilon > 0$, let $\mathcal{U} = \{x \in B : S_{a,K}(G,x) \geq 2\lambda, N(x) \leq \varepsilon\lambda\}$. Then, if B contains a point x_0 such that $S_{a,K}(G,x_0) \leq \lambda$, we can choose ε independent of λ and B so that $|\mathcal{U}| \leq \eta|B|$.

Proof. A moment's thouhgt will convince the reader that if r denotes the radius of B, then, for $x \in B$,

$$\Gamma_a(x) \subseteq \Gamma_a(x_0) \cup \{(y,t) : 0 \le at - 2r \le |x - y| \le at\}$$
$$\cup \left(\Gamma_a(x) \cap \{(y,t) : at \le 2r\} \right).$$

Thus, if $\tau \le \inf\{t : (y,t) \in K\}$, it follows that

$$\Gamma_a(x) \cap K \subseteq (\Gamma_a(x_0) \cap K) \cup \{(y,t) : 0 \le at - 2r \le |x - y| \le at\}$$
$$\cup \left(\Gamma_a(x) \cap \{(y,t) : \tau \le t \le 2r/a\} \right). \tag{20}$$

Now, $\mathcal{U}$ is closed, and if $x \in \mathcal{U}$, we have $G(y,t)^2 \le \varepsilon^2 \lambda^2$ for $|x - y| \le at$. Thus, integrating over the sets described in (20) above we find that

$$S_{a,K}(G,x)^2 \le S_{a,K}(G,x_0)^2 + (\varepsilon\lambda)^2 \frac{1}{v} \int_{2r/a}^{\infty} \int_{\{at - 2r \le |x-y| \le at\}} (at)^{-n} \, dy \frac{dt}{t}$$
$$+ \frac{1}{v} \int_{\tau}^{2r/a} \int_{|x-y| \le at} G(y,t)^2 (at)^{-n} \, dy \frac{dt}{t}.$$

But, since the first integral on the right-hand side above is a constant that depends only on the dimension n, and since $S_{a,K}(G,x_0) \le \lambda$, for $x \in \mathcal{U}$ we have

$$S_{a,K}(G,x)^2 \le (1 + c\varepsilon)^2 \lambda^2 + \frac{1}{v} \int_{\tau}^{2r/a} \int_{|x-y| \le at} G(y,t)^2 (at)^{-n} \, dy \frac{dt}{t}.$$

Integrating now over $\mathcal{U}$, and observing that

$$\frac{1}{v} \int_{\mathcal{U}} \chi(|x - y|/at)(at)^{-n} \, dx \le 1,$$

and that the above integral vanishes if the distance of y to $\mathcal{U}$ exceeds at, and that therefore this integral is majorized by the function $\Omega(y,at)$ associated with the complement of $\mathcal{U}$ as in Lemma 8, we obtain

$$\int_{\mathcal{U}} S_{a,K}(G,x)^2 \, dx \le |\mathcal{U}|(1 + c\varepsilon^2)\lambda^2 + \int_{\tau}^{2r/a} \int_{R^n} G(x,t)^2 \Omega(x,at) \, dx \frac{dt}{t}.$$

Moreover, since $S_{a,K}(G,x)^2 \ge 2\lambda$ on $\mathcal{U}$, the left-hand side of the preceding inequality is not less than $4\lambda^2 |\mathcal{U}|$. As for the right-hand side, it can be estimated by using Lemma 9. From the definition of $N(x)$ and $\mathcal{U}$ we see that $|F|, G$ and $|F_\psi|$ do not exceed $\varepsilon\lambda$ on the support of $\Omega_1(x,at)$, so that by Lemma 9 we obtain

$$\int_{\tau}^{2r/a} \int_{R^n} G(x,t)^2 \Omega(x,at) \, dx \frac{dt}{t}$$
$$\le \frac{3}{2}(\varepsilon\lambda)^2 \int_{R^n} \left(\Omega_1(x,a\tau) + \int_{\tau}^{2r/a} \frac{\partial}{\partial t} \Omega_1(x,at) \, dt \right) dx$$
$$+ (4 + c)(\varepsilon\lambda)^2 \int_{R^n} \int_{\tau}^{2r/a} \frac{\partial}{\partial t} \Omega_1(x,at) \, dt dx.$$

Furthermore, since $\Omega_1(x, at)$ vanishes if $t \le 2r/a$ and the distance between x and B is larger than $2r$, this last expression does not exceed $c_1(\varepsilon\lambda)^2|B|$, where c_1 is a constant. Combining these observations we get

$$4\lambda^2|\mathcal{U}| \le (1 + c\varepsilon^2)\lambda^2|\mathcal{U}| + c_1(\varepsilon\lambda)^2|B|,$$

from where the desired estimate follows by taking ε sufficiently small. ∎

A result of similar nature that is also useful in the sequel is

Lemma 11. Suppose F is as in Lemma 9, and let $B = B(x_0, r)$ be a fixed ball. Then, given $0 < s_0 < 1$, there exist constants $\bar{a} > 0$ and $0 < s_1 < 1$ such that for all $\lambda > 0$,

$$|\{y \in B : M_2^{0,2r}(N, y) > \bar{a}\lambda\}| < s_1|B|,$$
$$\text{implies} \quad |\{y \in B : S_1^{0,r}(G, y) > \lambda\}| < s_0|B|.$$

Proof. As the proof follows essentially that of the preceding material, we only sketch it here. Let $\bar{\eta}(t)$ be a nonincreasing positive $C_0^\infty(R^+)$ function equal to 1 for $0 < t \le 1$, and equal to 0 for $t \ge 2$, and set $\eta(x) = \bar{\eta}(|x - x_0|/r)$. Let $\mathcal{O}$, as in Lemma 10, be the set where $M_a^{0,2r}(N) > \bar{a}\lambda$, where the constant $\bar{a}$ is yet to be chosen. As above we see that, with Ω associated to $\mathcal{O}$ as in Lemma 8,

$$\int_{t_1}^{t_2} \int_{R^n} G(y, t)^2 \Omega_2(y, t)\, dy \frac{dt}{t} \le c\bar{a}^2\lambda^2 r^n, \quad 0 < t_1 < t_2 < r.$$

Moreover, since

$$\int_{\mathcal{O}^c \cap B} S_1^{0,r}(G, y)^2\, dy = \int_0^r \int_{\mathcal{O}^c \cap B} G(y, t)^2 \Omega(y, t)\, dy \frac{dt}{t}$$
$$\le \int_0^r \int_{R^n} G(y, t)^2 \Omega(y, t)\eta(y)\, dy \frac{dt}{t},$$

from the above estimate, upon letting $t_2 \to 0$ and $t_1 \to r$, it follows that

$$\int_{\mathcal{O} \cap B} S_1^{0,r}(G, y)^2\, dy \le c(\bar{a}\lambda)^2 r^n. \tag{21}$$

Consequently, $|\{y \in B : S_1^{0,r}(G, y) > \lambda\}|$ is equal to

$$|\{y \in B \cap \mathcal{O} : S_1^{0,r}(G, y) > \lambda\}| + |\{y \in B \cap \mathcal{O}^c : S_1^{0,r}(G, y) > \lambda\}|$$
$$\le |\mathcal{O} \cap B| + c_1 \bar{a}^2|B|,$$

where to estimate the measure of the second set we applied Chebychev's inequality to (21).

Now fix $\bar{a}$ so that $c_1\bar{a}^2 < s_0/2$, and then choose s_1 above so that $s_1 < s_0/2$; since $|\mathcal{O} \cap B| < s_1|B|$, this gives the desired estimate. ∎

Proof of Theorem 2. Since $M_a(F, x) \in L_w^p(R^n)$, from the results in Chapter V it follows that, with the notation of Lemmas 9 and 10, also $M_a(N, x) \in L_w^p(R^n)$, and $\|M_a(N)\|_{L_w^p} \leq c\|M_a(F)\|_{L_w^p}$. Moreover, since $w \in A_\infty$, from Lemma 8 it follows that

$$\nu(\{y \in B : S_{a,K}(G, y) > 2\lambda, N(y) \leq \varepsilon\lambda\}) \leq c\eta\nu(B), \quad \text{all } B. \tag{22}$$

For each x in $\{S_{a,K}(G) > \lambda\}$, let $B(x)$ be the largest ball centered at x whose interior is contained in $\{S_{a,K}(G) > \lambda\}$. There exists, then, a pairwise disjoint countable family $\{B(x_i)\}$, say, of such balls such that $\{S_{a,K}(G) > \lambda\} \subseteq \bigcup_i B^1(x_i)$. Since these balls are pairwise disjoint, their interiors are contained in $\{S_{a,K}(G) > \lambda\}$ and ν is doubling, it follows that

$$\nu(\{S_{a,K}(G) > \lambda\}) \geq \sum_i \nu(\overline{B}(x_i)) \geq c_1 \sum_i \nu(B^1(x_i)), \tag{23}$$

where $\overline{B}(x_i)$ denotes the closure of the ball $B(x_i)$. Each ball $\overline{B}(x_i)$ contains a point where $S_{a,K} \leq \lambda$. Therefore, by Lemma 10 and (22) above it follows that

$$\nu(\{y \in \overline{B}_i : S_{a,K}(G, y) > 2\lambda, N(y) \leq \varepsilon\lambda\}) \leq c\eta\nu(\overline{B}_i).$$

Whence, since $\{S_{a,K} > 2\lambda, N \leq \varepsilon\lambda\} \subseteq \{S_{a,K} > \lambda\} \subseteq \bigcup_i \overline{B}_i$, summing over i we get

$$\nu(\{S_{a,K}(G) > 2\lambda, N \leq \varepsilon\lambda\}) \leq c_2\eta\nu(\{S_{a,K} > \lambda\}),$$

which in turn implies that

$$\nu(\{S_{a,K}(G) > 2\lambda\}) \leq c_2\eta\nu(\{S_{a,K}(G) > \lambda\}) + \nu(\{N > \varepsilon\lambda\}).$$

Integrating this inequality with respect to λ from 0 to ∞ with respect to the measure $d\lambda^p$, $0 < p < \infty$, we obtain

$$2^{-p}\|S_{a,K}(G)\|_{L_w^p}^p \leq c_2\eta\|S_{a,K}(G)\|_{L_w^p}^p + \varepsilon^{-p}\|N\|_{L_w^p}^p.$$

Now, according to the definition of $S_{a,K}(G)$, since K is compact and contained in $t > 0$, $S_{a,K}(G)$ is continuous and has compact support, and so $\|S_{a,K}(G)\|_{L_w^p} < \infty$. Thus, from the last inequality, upon choosing $c_2\eta < 2^{-p}$ we find that

$$\|S_{a,K}(G)\|_{L_w^p} \leq c\|N(F)\|_{L_w^p} \leq c\|M_b(F)\|_{L_w^p}.$$

Whence, letting K increase and exhaust the upper half-space, passing to the limit we obtain

$$\|S_a(G)\|_{L_w^p} = \|\lim S_{a,K}(G)\|_{L_w^p} \le c\|M_b(F)\|_{L_w^p}. \quad \blacksquare$$

Proof of Theorem 4. The left-hand side inequality follows since the multiplier operators T_i are bounded on $H_w^p(R^n)$, cf. Chapter XI. As for the right-hand side inequality, we invoke Theorem e in Chapter V, as it applies to the f_i's. Let $(x,t) \in R_+^{n+1}$, then for any $q > 0, N > 0$, we have

$$\nu(B(x,t))|(f_i * \phi_t)(x)|^q \le \liminf_{\varepsilon \to 0} \int_{R^n} (1 + (|x-y|/t))^{-Nq}|f_i * \phi_t(y)|^q w(y)\,dy;$$

here we used the fact that the Fourier transform is supported in a cone Γ_i. From this estimate it readily follows that

$$|(T_if * \phi_t)(x)|^q \le cM_\nu(|T_if|^q)(z), \quad \text{for } |x - z| < t.$$

Thus, if $F_{i,\phi}(x,t) = (T_i * \phi_t(x))$, we have $M_1(F_{i,\phi},x) \le cM_\nu(|T_if|^q)(x)^{1/q}$, and for $p > q$ it follows that

$$\|f\|_{H_w^p} \sim \|M_1(F_{i,\phi}\|_{L_w^p} \le c\sum_{i=1}^k \|M_1(F_{i,\phi})\|_{L_w^p}$$

$$\le c\sum_{i=1}^k \|M_\nu(|T_if|^q)\|_{L_w^{p/q}} \le c\sum_{i=1}^k \|T_if\|_{L_w^p}.$$

In the last step we used that, since $p/q > 1$, M_ν is bounded on $L_w^{p/q}(R^n)$. $\quad \blacksquare$

Proof of Theorem 5. First assume that $f \in H_w^p(R^n)$. Since $\hat{\mathcal{D}}_0$ is dense in $H_w^p(R^n)$, cf. Chapter VII, the operators T_i can be extended continuously to $H_w^p(R^n)$ as follows: Given $f \in H_w^p(R^n)$, let $\{f_m\} \subset \hat{\mathcal{D}}_0$ be a sequence that converges to f in $H_w^p(R^n)$, and set $g_i = \lim_{m\to\infty} T_if_m$. Then, if $g \in \hat{\mathcal{D}}_0$, we have $\langle T_if, g\rangle = \lim_{m\to\infty}\langle T_if_m, g\rangle = \lim_{m\to\infty}\langle f_m, T_ig\rangle = \langle f, T_ig\rangle$; thus, $\sum_{i=1}^k f_i = \sum_{i=1}^k \lim_{m\to\infty} T_if_m = \lim_{m\to\infty}\sum_{i=1}^k T_if_m = \lim_{m\to\infty} g_m = f$. Moreover, $\|T_if\|_{H_w^p} = \lim_{m\to\infty}\|T_if_m\|_{H_w^p} = c\lim_{m\to\infty}\|f_m\|_{H_w^p} = c\|f\|_{H_w^p}$. It then follows that

$$\liminf_{\varepsilon > 0} \|T_if * \phi_\varepsilon\|_{L_w^p} \le \sup_{\varepsilon > 0} \|T_if * \phi_\varepsilon\|_{L_w^p}$$

$$\le \|\sup_\varepsilon (T_i * \phi_\varepsilon)\|_{L_w^p} \le c\|T_if\|_{H_w^p} \le c\|f\|_{H_w^p}.$$

By summation over i, the left-hand side inequality follows.

Next assume that $\liminf_{\varepsilon>0} \|f_i * \phi_i\|_{L_w^p} < \infty$ for $i = 1, \ldots, k$. Pick $0 < q < p$, and let $\{\varepsilon_k\}$ be a sequence such that

$$\|f_i * \phi_{\varepsilon_k}\|_{L_w^p} \to \liminf_{\varepsilon\to 0} \|f_i * \phi_\varepsilon\|_{L_w^p},$$

and so that $|f_i * \phi_{\varepsilon_k}|^q$ converges weakly to a function g_i, say, in $L_w^{p/q}(R^n)$. Then,

$$\|g_i\|_{L_w^{p/q}} \leq \liminf_{\varepsilon\to 0} \|f_i * \phi_\varepsilon\|_{L_w^p}, \quad i = 1, \ldots, k.$$

Since the support of $\hat{f}_i$ lies in a cone Γ_i it follows that

$$\nu(B(x,t))|(f_i * \phi_\varepsilon)(x)|^q$$
$$\leq c \liminf_{\varepsilon\to 0} \int_{R^n} |(f_i * \phi_\varepsilon)(y)|^q (1 + (|x-y|/t))^{-Nq} w(y)\,dy$$
$$\leq c \lim_{k\to\infty} \int_{R^n} |(f_i * \phi_{\varepsilon_k})(y)|^q (1 + (|x-y|/t))^{-Nq} w(y)\,dy.$$

Further, since $(1 + (|x-y|/t))^{-Nq} \in L_w^{p/q}(R^n)$, by the weak convergence,

$$\nu(B(x,t))|(f_i * \phi_t)(x)|^q \leq c \int_{R^n} |g_i(y)|(1 + (|x-y|/t))^{-Nq} w(y)\,dy,$$

and consequently,

$$|f_i * \phi_t(x)|^q \leq c M_\nu g_i(z) \quad \text{for} \quad |x-z| < t.$$

Thus, $M_1(F_{i,\phi}, x) \leq c(M_\nu g_i(x))^{1/q}$, and

$$\|f\|_{H_w^p} \sim \|M_1(F_\phi)\|_{L_w^p} \leq \sum_{i=1}^k \|M_1(F_{i,\phi})\|_{L_w^p}$$
$$\leq c \sum_{i=1}^k \|M_\nu g_i\|_{L_w^p} \leq c \sum_{i=1}^k \|g_i\|_{L_w^{p/q}}^{p/q} = c \sum_{i=1}^k \liminf_{\varepsilon\to 0} \|f_i * \phi_\varepsilon\|_{L_w^p}.$$

This gives the right-hand side inequality, and completes the proof. ∎

Now, in case of the nonisotropic dilations, each of the cones $\Gamma_{i\pm}$, $i = 1, \ldots, n$, lies strictly on one side of the hyperplane $x_i = 0$, and the arguments in the proof of Theorem 3 in Chapter V apply to tempered distributions supported in such cones. The above proof can then be carried out with no modifications.

Sources and Remarks. In the context of analytic Hardy spaces the results covered in this chapter are classical, cf. A. Zygmund [1959]. In the unweighted case, the corresponding results were proved by C. Fefferman and E. Stein [1972], and D. Burholder and R. Gundy [1972]. The method we use here, to wit, that of good-lambda inequalities, originates in the work of D. Burkholder and R. Gundy, and was carried out in the context of weighted spaces by R. Gundy and R. Wheeden [1974]. The use of the heat kernel originates in the work of A. Calderón and A. Torchinsky [1975]. The observation of E. Stein that the norm of functions in $H^1(R^n)$ can be computed in terms of the $L^1(R^n)$ norm of the function and its Riesz transforms is also true in the weightes spaces, cf. R. Wheeden [1976]. A similar result is true for the theory of E. Stein and G. Weiss in terms of systems of conjugate harmonic functions in the upper half-space.

CHAPTER

VII

A Dense Class

An important feature of the weighted Hardy spaces is that they have a reasonable dense class of functions, namely $\hat{\mathcal{D}}_0$. Using this fact we show, in the next chapter, that finite linear combinations of atoms are also dense.

The basic result in this chapter is, then,

Theorem 1. Suppose ν is a doubling weighted measure with respect to the Lebesgue measure of R^n with weight w. Then, $\hat{\mathcal{D}}_0$ is dense in $H^p_w(R^n)$, $0 < p < \infty$.

The proof of this result is achieved in several steps. A step we single out is the existence of nontangential boundary values.

Theorem 2. Let $f \in \mathcal{S}'(R^n)$ be a tempered distribution, and $\phi \in \mathcal{S}(R^n)$ a Schwartz function with integral 1. Suppose that for some $\lambda > 0$, $N_\lambda(F_\phi, x)$ is finite a.e. in a set $\mathcal{E}$. Then the nontangential limit

$$\lim_{\substack{|x-y| < at \\ t \to 0}} F_\phi(y, t) = g(x),$$

exists a.e. in $\mathcal{E}$. Furthermore, $g(x)$ is independent of ϕ and $a > 0$.

Using estimates established in Chapter V we reduce the proof of Therem 2 to that of

Lemma 3. Let $H(x, t)$ be a function in R^{n+1}_+ which satisfies the heat equation, and assume that $M_{2a}(H, x)$ is finite for almost all x in a set $\mathcal{E}$. Then the nontangential limit

$$\lim_{\substack{|x-y|<at \\ t\to 0}} H(y,t) = h(x)$$

exists a.e. in $\mathcal{E}$.

Proof that Lemma 3 implies Theorem 2. Let $H(x,t) = f * \psi_t(x)$, where $\psi_t(x) = t^{-n} e^{-\pi(|x|/t)^2}$. From Theorem 8 in Chapter V it follows that $M_{2a}(H,x) \leq c_{a,\lambda} N_\lambda(F_\phi, x)$ for any $a > 0$. Thus, under our hypothesis, $M_{2a}(H,x)$ is finite a.e. on $\mathcal{E}$ and, by Lemma 3, we conclude that the nontangential limit

$$\lim_{\substack{|x-y|<at \\ t\to 0}} H(y,t) = g(x)$$

exists a.e. for x in $\mathcal{E}$ for all $a > 0$. Thus, the limit in Lemma 3 exists for almost all x in $\mathcal{E}$ for all $a > 0$. Combining results in Chapters IV and V it readily follows that $N_\lambda(H,x) \leq c_\lambda N_\lambda(F_\phi, x)$, and consequently, $N_\lambda(H,x)$ is also finite for almost all $x \in \mathcal{E}$. Let $x_0 \in \mathcal{E}$ be a point where $N_\lambda(H,x_0)$ is finite and the nontangential limit in Lemma 3 exists for all $a > 0$. It suffices to show that F_ϕ has a nontangential limit at x_0; since this is precisely the conclusion of Theorem 13 in Chapter V, we are done.

Proof of Lemma 3. Let $\tilde{M}_a^{0,\delta}(F,x)$ be the maximal function introduced in the proof of Theorem 3 in Chapter VI. To arrive at the conclusion we prove that $\lim_{\delta\to 0} \tilde{M}_a^{0,\delta}(F,x) = 0$ for almost all $x \in \mathcal{E}$. Let $\mathcal{E}_1$ be the subset of $\mathcal{E}$ where $\lim_{\delta\to 0} \tilde{M}_a^{0,\delta}(F,x) > \varepsilon$, for some $\varepsilon > 0$; we will show that $|\mathcal{E}_1| = 0$. To achieve this, by the Lebesgue density theorem, it suffices to find $0 < \varepsilon_0 < 1$ such that

$$|\mathcal{E}_1 \cap B| < (1-\varepsilon_0)|B|, \quad \text{all balls } B.$$

Fix a ball B and pick $0 < \varepsilon_1 < 1$, ε_1 is yet to be chosen, independently of B. Since $\mathcal{E}_1 \subseteq \mathcal{E}$, if $|\mathcal{E} \cap B| < (1-\varepsilon_1)|B|$, the desired estimate is trivial, so, assume instead that $|\mathcal{E} \cap B| > (1-\varepsilon_1)|B|$. Since the nontangential maximal function is finite on the set $\mathcal{E}$, we can find a large number $K_B > 0$, say, depending on B, such that the nontangential maximal function is less than or equal to K_B on a set $\mathcal{E}_2$ with $|\mathcal{E}_2 \cap B| \geq (1-2\varepsilon_1)|B|$. By Lemma 11 in Chapter VI, we conclude that the Lusin function truncated at the height $h(B) = $ radius of B, is bounded by $c_1 K_B$ on a set $\mathcal{E}_3$ with $|\mathcal{E}_3 \cap B| > (1-\varepsilon_2)|B|$, where $1 > \varepsilon_2 = \varepsilon_2(\varepsilon_1, c_1) > 0$ can be chosen arbitrarily small provided we pick ε_1 sufficiently small and c_1 large enough, independently of B. Further, since the Lusin function is given by an integral with nonnegative integrand, it can be made arbitrarily small by means of a small truncation, provided it is finite for some truncation. Thus, we can find a set $\mathcal{E}_4 \subseteq \mathcal{E}_3$ with $|\mathcal{E}_4 \cap B| > (1-2\varepsilon_2)|B|$, a constant $\varepsilon_4 > 0$ and $\delta_1 = \delta_1(\varepsilon_4, B)$, such that

$$S_a^{0,\delta_1}(G,x) \leq \varepsilon_4, \quad x \in \mathcal{E}_4.$$

Now we cover B with a collection $\{B_i\}$ of balls with radius δ_1 so that at most finitely many of them, say N, overlap. First consider those B_i's such that $|\mathcal{E}_4 \cap B| \leq (1 - \sqrt{2\varepsilon_2})|B_i|$; the sum of the measures of these balls does not exceed $N\sqrt{2\varepsilon_2}|B|$. On the other hand, if $|\mathcal{E}_4 \cap B| > (1 - \sqrt{2\varepsilon_2})|B|$, by Lemma 6 in Chapter VI, it follows that $\tilde{M}_a^{0,\delta_1} < c_2\varepsilon_4$ on a set $\mathcal{E}_5$ with $|\mathcal{E}_5 \cap B_i| > (1 - \varepsilon_5)|B_i|$, where ε_5 depends on c_2 and ε_2. Whence, we have

$$|B \setminus \mathcal{E}_5| \leq N\sqrt{2\varepsilon_2}\,|B| + \varepsilon_5 \sum_i |B_i| \leq c(\sqrt{2\varepsilon_2} + \varepsilon_5)|B|\,.$$

Now observe that we can get $c\varepsilon_5 \leq 1/3$ by choosing c_2 large and ε_2 small, and then we can also get $c\sqrt{2\varepsilon_2} \leq 1/3$ by picking ε_1 small; we also want $\varepsilon_1 \leq 1/3$. Finally, we need $c_2\varepsilon_4 < \varepsilon$. Then it follows that $|B \setminus \mathcal{E}_5| < (2/3)|B|$, and since $\tilde{M}_a^{0,\delta_1} < \varepsilon$ on the set $\mathcal{E}_5$, we conclude that $\mathcal{E}_1 \cap B \subseteq B \setminus \mathcal{E}_5$. Whence,

$$|\mathcal{E}_1 \cap B| \leq (1 - \varepsilon_0)|B|\,, \quad \text{with} \quad \varepsilon_0 = 1/3\,. \quad \blacksquare$$

The proof of Theorem 1 proceeds now in steps.

Step 1. The class of functions whose Fourier transform coincides with a distribution with compact support not containing the origin is dense in $H_w^p(R^n)$, $0 < p < \infty$.

Step 2. $\hat{\mathcal{D}}_0$ is dense in $H_w^p(R^n)$, $0 < p < \infty$.

Proof of Step 1. Let $f \in H_w^p(R^n)$, and suppose $\phi \in \mathcal{S}(R^n)$ has the property that $\hat{\phi}(\xi) = 1$ for $|\xi| \leq 1$, and $\hat{\phi}(\xi) = 0$ for $|\xi| \geq 2$; we want to prove that

$$\|f - F_\phi(\cdot,s) + F_\phi(\cdot,1/s)\|_{H_w^p} \to 0\,, \quad \text{as } s \to 0\,.$$

We do so by showing that

$$\|f - F_\phi(\cdot,s)\|_{H_w^p}\,, \quad \|F_\phi(\cdot,1/s)\|_{H_w^p} \to 0\,, \quad \text{as } s \to 0\,.$$

We begin by proving two lemmas.

Lemma 4. Suppose that $G(y,t)$ has a nontangential limit L at the point $x \in R^n$ through cones $|x - y| < at$ for every $a > 0$, and that

$$M_\lambda(G,x,t) = \sup_{y \in R^n} |G(y,t)|(1 + (|x - y|/t))^{-\lambda} < \infty$$

for $0 < t \leq t_0$. Then, if ψ is a Schwartz function with integral equal to 1, $(G(\cdot,t) * \psi_s)(y)$ converges nontangentially to L as $s,t \to 0$, $0 < s < t$, through the cones $|x - y| < a't$ for every $a' > 0$.

Proof. With no loss of generality suppose that $a' = 1$. Observe that for $|x-y| < t$ we have

$$|(G(\cdot,t) * \psi_s)(y) - L| \leq \int_{R^n} |G(y-z,t) - L| \, |\psi_s(z)| \, dz$$

$$\leq \left(\int_{\{|z|<\delta t\}} + \int_{\{|z|\geq \delta t\}} \right) = I + J,$$

say. Now, J does not exceed

$$c \int_{\{|z|\geq \delta t\}} |\psi_s(z)|(1 + (|x - y + z|/t))^\lambda \, dz$$

$$\times \sup_{z\in R^n} \left(|G(y - z,t) - L|(1 + (|x - y + z|/t))^{-\lambda} \right)$$

$$\leq c \left(\int_{\{|z|\geq \delta t\}} |\psi_s(z)| \, (1 + (|x - y|/t) + (|z|/s))^\lambda \, dz \right) (M_\lambda(G, x, t) + |L|)$$

$$\leq c \left(\int_{\{|z|\geq \delta t/s\}} |\psi(z)|(2 + |z|)^\lambda \, dz \right) (M_\lambda(G, x, t) + |L|),$$

and this expression goes to 0 as $\delta \to \infty$. Also,

$$I \leq \|\psi\|_{L^1} \sup_{|x-z|\leq(\delta+1)t} |G(z,t) - L|,$$

which goes to 0 for every fixed δ as $t \to 0$. ∎

Lemma 5. For every Schwartz function ψ we have

$$|(G(\cdot,t) * \psi_s)(y)| \leq cM_\lambda(G, x, t), \quad \text{whenever} \quad |x - y| < t, s < t.$$

Proof. We estimate the left-hand side above by

$$\int_{R^n} |G(y - z,t)|(1 + (|x - y + z|/t))^{-\lambda}|\psi_s(z)|(1 + (|x - y + z|/t))^\lambda \, dz$$

$$\leq M_\lambda(G, x, t) \int_{R^n} |\psi_s(z)|(1 + (|x - y|/t) + (|z|/s))^\lambda \, dz$$

$$\leq cM_\lambda(G, x, t) \int_{R^n} |\psi(z)|(1 + |z|)^\lambda \, dz. \quad ∎$$

Proof of Step 1, continued. Let ψ be a Schwartz function with integral 1. Setting for convenience $F_\phi = F$, we have

$$(f * \phi_s * \psi_t)(y) = (F(\cdot,s) * \psi_t)(y) = (F_\psi(\cdot,t) * \phi_s)(y).$$

Thus, from Lemma 5 it follows that for $|x - y| < t$,

$$|(F(\cdot, s) * \psi_t)(y)| \le cM_\lambda(F, x, s), \quad \text{or} \quad \le cM_\lambda(F_\psi, x, t),$$

depending whether $s \le t$ or $t \le s$. Consequently, for all $s > 0$ we have

$$M_1((F(\cdot, s))_\psi, x) \le c(N_\lambda(F, x) + N_\lambda(F_\psi, x)).$$

The functions on right-hand side above are in $L_w^p(R^n)$ provided that λ is sufficiently large, depending on p and on the doubling exponent of ν. Moreover, since

$$M_1((F - F(\cdot, s))_\psi, x) \le M_1(F, x) + M_1(F(\cdot, s), x),$$

to verify that $\|F - F(\cdot, s)\|_{H_w^p} \to 0$ as $s \to 0$, by the Lebesgue dominated convergence theorem, it suffices to show that $M_1((F - F(\cdot, s))_\psi, x) \to 0$ as $s \to 0$ a.e. in x; and similarly for $M_1((F(\cdot, 1/s))_\psi, x)$.

Note that $M_\lambda(F_\psi, x, t) \to 0$ as $t \to \infty$. Indeed, there is an absolute constant c such that $M_\lambda(F_\psi, x, t) \le cM_\lambda(F_\psi, y, t)$ whenever $|y - x| < t$. Also $M_\lambda(F_\psi, y, t) \le N_\lambda(F_\psi, y)$. Whence,

$$\nu(B(x, t))M_\lambda(F_\psi, x, t)^p \le c\int_{B(x,t)} N_\lambda(F_\psi, y)^p w(y)\, dy \le c\|f\|_{H_w^p}^p,$$

and since $\lim_{t \to \infty} \nu(B(x, t)) = \infty$, as asserted, $\lim_{t \to \infty} M_\lambda(F_\psi, x, t) = 0$; similarly, $\lim_{s \to \infty} M_\lambda(F, x, s) = 0$.

As for the other statements, note that from Lemma 5 it follows that

$$M_1(F(\cdot, 1/s)_\psi, x) \le cM_\lambda(F_\psi, x, 1/s) + c \sup_{t > 1/s} M_\lambda(F, x, t),$$

and consequently, $M_1((F(\cdot, 1/s)_\psi, x) \to 0$ a.e. as $s \to \infty$; this is one of the statements we wanted to show. To prove the other, note that $M_1((F - F(\cdot, s))_\psi, x)$ may be estimated by

$$M_1^{0,\delta}((F - F(\cdot, s))_\psi) + M_1^{\delta,T}((F - F(\cdot, s))_\psi, x) + M_1^{T,\infty}(F_\psi, x)$$
$$+ M_1^{T,\infty}((F(\cdot, s))_\psi, x) = A + B + C + D,$$

say. Provided that $s < T$, C and D are bounded by $c\sup_{t > T} M_\lambda(F, x, t)$, and consequently, these terms go to 0 as $T \to \infty$. By Lemma 4, $A \to 0$ for almost all x, as $s, \delta \to 0$, if $s < \delta$. To complete the proof we show that $B \to 0$ for almost all x as $s \to 0$ with δ and T fixed, $\delta < T$. In other words, we verify that $F(y, t) - (F(\cdot, t) * \psi_s)(y)$ converges uniformly to 0 in the set $|x - y| < t$, $\delta < t < T$. Well, we have that

$$|F(y, t) - (F(\cdot, t) * \psi_s)(y)| \le \int_{R^n} |\psi_s(z)|\, |F(y - z, t) - F(y, t)|\, dz$$

$$= \left(\int_{\{|z| < rs\}} + \int_{\{|z| \ge rs\}}\right) = B_0 + B_1,$$

say. The following estimates then hold:

$$B_0 \leq \sup_{|z|<rs} |F(y-z,t) - F(y,t)| \, \|\psi\|_{L^1} \,,$$

and

$$B_1 \leq \sup_{\delta<t<T} M_\lambda(F,x,t) \int_{\{|z|>r\}} |\psi(z)|(1+|z|)^\lambda \, dz \,.$$

B_1 goes to 0 as $r \to \infty$, and, since $F(y,t)$ is uniformly continuous in the set $|x-y| < 2t, \delta < t < T$, it follows that $B_0 \to 0$ uniformly in $|x-y| < t, \delta < t < T$, as $s \to 0$ for each fixed r. We have thus completed the proof of Step 1.

Proof of Step 2. We may assume that f is actually a function whose Fourier transform $\hat{f}$ is a distribution with compact support that does not contain the origin. Let η be a Schwartz function with $\eta(0) = 1$ and such that $\hat{\eta}$ has compact support, and let $f^{(s)}(x) = f(x)\eta(sx)$. Then the distribution $\hat{f}^{(s)} = \hat{f} * \hat{\eta}_s$, where $\hat{\eta}_s(\xi) = s^{-n}\hat{\eta}(\xi/s)$, is compactly supported and it vanishes near the origin for sufficiently small s, say $s < s_0$. Further, since $\hat{f}$ is a distribution with compact support and $|\eta| < c$, we have $|f^{(s)}(x)| \leq c_1|f(x)| \leq c(1+|x|)^M$ for some $M > 0$.

We will show that $\|f^{(s)} - f\|_{H_w^p} \to 0$ as $s \to 0$. Let $\phi \in \hat{\mathcal{D}}$ be a function with integral equal to 1, and consider

$$H^{(s)}(x,t) = ((f - f^{(s)}) * \phi_t)(x) = \int_{R^n} f(x-y)(1 - \eta(s(x-y))\phi_t(y) \, dy \,.$$

Since $\hat{\phi}$ has compact support and $\hat{f} - \hat{f}^{(s)}$ vanishes in a neighborhoord of the origin for $s < s_0$, it follows that $H^{(s)}(x,t) = 0$ provided that $s < s_0$ and t is sufficiently large, say $t > t_0$; we want to estimate $\|M_1(H^{(s)}, \cdot)\|_{L_w^p}$.

By Taylor's formula we have

$$1 - \eta(s(x-y)) = - \sum_{0<|\sigma|<m} \eta_\sigma(sx)s^{|\sigma|}\frac{(-y)^\sigma}{\sigma!} + R(sx,sy) \,,$$

where $\eta_\sigma = D^\sigma \eta$, and $R(x,y)$ satisfies the estimates

$$|R(x,y)| \leq c_m|y|^m$$

and

$$|R(x,y)| \leq c_{m,k}|y|^m(1+|x|)^{-k} \,, \quad |y| \leq |x|/2 \,,$$

the second inequality being valid for each k with an appropriate constant $c_{m,k}$. Whence, substituting this expression in the integral representation of $H^{(s)}(x,t)$

we obtain that it equals

$$\sum_{0<|\sigma|<m} (st)^{|\sigma|}\frac{1}{\sigma!}\eta_\sigma(sx)\int_{R^n}(-y/t)^\sigma \phi_t(y)f(x-y)\,dy$$

$$+\int_{R^n} R(sx,sy)\phi_t(y)f(x-y)\,dy = H_1^{(s)}(x,t)+H_2^{(s)}(x,t),$$

say. Put now $\phi^\sigma(y)=(-y)^\sigma\phi(y)$. Since, as may be checked directly, $f^{(s)}$ is in $H_w^p(R^n)$, we have $\|M_1(\eta_\sigma(s\cdot)(f^{(s)}*\phi_t^\sigma))\|_{L_w^p} < \infty$ for all σ. As each term in the sum defining $H_1^{(s)}(x,t)$ has a positive power of (st) as a factor, and since we may assume that $t \le t_0$, it readily follows that $\|M_1(H_1^{(s)})\|_{L_w^p} \to 0$ as $s \to 0$.

It thus only remains to bound $|H_2^{(s)}(x,t)|$. From the above estimates for f and $R(x,y)$, it does not exceed

$$c(1+s|x|)^{-k}\int_{R^n}(s|y|)^m(1+|x-y|)^M|\phi_t(y)|\,dy$$

$$+c\int_{\{|y|>|x|/2\}}(s|y|)^m(1+|x-y|)^M|\phi_t(y)|\,dy\,.$$

Moreover, since $(1+|x-y|)^M \le c(1+|x|)^M(1+|y|)^M$ and since $|\phi_t(y)| \le ct^{-n}(1+(|y|/t))^{-\ell}$ for any ℓ, with c depending, of course, on ℓ, it follows that $H_2^{(s)}(x,t)$ does not exceed,

$$cs^m(1+s|x|)^{-k}(1+|x|)^M\int_{R^n}|y|^m(1+|y|)^M t^{-n}(1+(|y|/t))^{-\ell}\,dy$$

$$+cs^m(1+|x|)^M\int_{\{|y|>|x|/2\}}|y|^M(1+|y|)^M t^{-n}(1+(|y|/t))^{-\ell}\,dy\,.$$

Consequently, for $0<s<s_0$ and $0<t<t_0$, the above expression is dominated by

$$ct^{(m+n)-n}s^{m-k}(1+|x|)^{M-k}\int_{R^n}(|y|/t)^m(1+(|y|/t))^{M-\ell}t^{-n}\,dy$$

$$+ct^{(m+n)-n}s^m(1+|x|)^M\int_{\{|y|>|x|/2\}}(|y|/t)^m(1+(|y|/t))^{M-\ell}t^{-n}\,dy$$

$$\le ct^m s^{m-k}(1+|x|)^{M-k}\int_{R^n}|y|^m(1+|y|)^{M-\ell}\,dy$$

$$+ct^m s^m(1+|x|)^M\int_{\{|y|>t_0^{-1}|x|/2\}}(1+|y|)^{M+m-\ell}\,dy\,.$$

Whence, if $M+m-\ell < -n$ it follows that

$$|H_2^{(s)}(x,t)| \le ct^m\left(s^{m-k}(1+|x|)^{M-k}+s^m(1+|x|)^{2M+m+n-\ell}\right).$$

Now, given $\ell' > 0$, first pick k so large that $M - k < -\ell'$, and then choose $m > 0$ so large that $m - k > 1$, and ℓ so large that $2M + m + n - \ell < -\ell'$. Using the above estimate for $H_2^{(s)}$ and the above choice for the parameters we get

$$M_1(H_2^{(s)}, x) = M_1^{0,t_0}(H_2^{(s)}, x) \le cs(1 + |x|)^{-\ell'}, \quad 0 < s \le s_0.$$

Thus, if ℓ' is large enough, $\int_{R^n}(1 + |x|)^{-\ell' p} w(x)\,dx < \infty$, and consequently, $\|M_1(H_2^{(s)})\|_{L_w^p} \le cs$ goes to 0 as $s \to 0$. This completes the proof of Step 2 and that of Theorem 1. ∎

Sources and Remarks. That $\hat{\mathcal{D}}_0$ is dense in the unweighted Hardy spaces was first established by A. Calderón and A. Torchinsky [1975], the main ideas for Step 2 are in their work; Stein [1970] had already proved the important particular case $p = 1$. The classes of tempered distributions f with $M_1^{0,1}(F_\phi, x) \in L^p(R^n)$, even locally, are of interest, and they have been studied by D. Goldberg [1979] and B. Marshall [1980]; the latter author also discusses nontangential convergence.

VIII

The Atomic Decomposition

The identification by C. Fefferman of the dual of $H^1(R^n)$ as the space of functions of bounded mean oscillation, $BMO(R^n)$, led R. Coifman to observe that functions in $H^1(R)$ can be decomposed into sums of simpler components called atoms.

The purpose of this chapter is to show that if ν is a doubling weighted measure with respect to the Lebesgue measure in R^n with weight w, then there is an atomic decomposition for elements $f \in H_w^p(R^n)$, $0 < p < \infty$.

Assume that $1 < q \le \infty$, and that N is a nonnegative integer. A (q, N) atom $a(x)$ is an $L^q(R^n)$ function with compact support and with vanishing moments of order less than or equal to N. More precisely, it is a function which satisfies the following three properties:

1. $\operatorname{supp} a \subseteq$ ball B.
2. $\|a\|_{L^q} \le \|\chi_B\|_{L^q}$.
3. $\int_{R^n} a(x) x^\alpha \, dx = 0$ for all $|\alpha| \le N$.

We then have

Theorem 1. Suppose ν is a doubling weighted measure with respect to the Lebesgue measure in R^n, $\nu \in D_d$, with weight w, and that f belongs to $H_w^p(R^n)$, $0 < p < \infty$. Then there is an integer $N(p, w)$ with the following property: Given an integer $N \ge N(p, w)$, we can find a sequence $\{\lambda_k\}$ of positive coefficients and a sequence $\{a_k\}$ of (∞, N) atoms with support contained in balls $\{B_k\}$ respectively, such that the sum $\sum_k \lambda_k a_k$ converges unconditionally, i.e., independently of the order the summands are taken, to f, both in the sense of distributions and in the

$H_w^p(R^n)$- norm. Moreover,

$$\left\| \sum_k \lambda_k^s \chi_{B_k} \right\|_{L_w^{p/s}} \leq c \|f\|_{H_w^p}, \quad 0 < s < \infty,$$

with the constant $c = c(p, d, s)$.

Conversely, if we have a sequence $\{a_k\}$ of (∞, N) atoms, and a sequence $\{\lambda_k\}$ of positive scalars such that

$$\left\| \sum_k \lambda_k^s \chi_{B_k} \right\|_{L_w^{p/s}} < \infty, \quad \text{for some } 0 < s \leq 1,$$

then $\sum_k \lambda_k a_k$ converges unconditionally in the sense of distributions and in the $H_w^p(R^n)$-norm to an element $f \in H_w^p(R^n)$ with

$$\|f\|_{H_w^p} \leq c \left\| \sum_k \lambda_k^s \chi_{B_k} \right\|_{L_w^{p/s}},$$

where $c = c(p, d, s)$. If in addition $w \in RH_{(q/p)'}$, with $q > \max(1, p)$, the result is also true for (q, N) atoms.

A word about the minimum amount of moments required. let $[\cdot]$ denote the "greatest integer less than or equal to" function. We then have

$$N(p, w) = \begin{cases} [(d/p) - n] & \text{if } 0 < p \leq 1 \\ [d - n] & \text{if } 1 < p < \infty. \end{cases}$$

On the other hand, if also $w \in A_{p_0}$, $p_o > p$, we have

$$N(p, w) = \left[\left(1 - \frac{p-1}{p_0 - 1} \right)(d - n)/p \right].$$

Finally, if $w \in A_p$, and $p > 1$, we set

$$N(p, w) = -1,$$

i.e., no moment condition is required of the atoms, and, by Theorem 1 in Chapter VI, $H_w^p(R^n)$ can be identified with $L_w^p(R^n)$.

We begin by proving some preliminary results.

Lemma 2. If the sequence $\{f_j\}$ of distributions in $H_w^p(R^n)$ converges to a distribution $f \in H_w^p(R^n)$ in the $H_w^p(R^n)$-norm, then it also converges to f in the sense of distributions.

Proof. Let B denote the unit ball of R^n, and $\phi \in \mathcal{S}(R^n)$. Then there is a positive constant c_ϕ which depends on the Schwartz norms of ϕ, such that if $\langle f, \phi \rangle$ denotes the evaluation of f at the test function ϕ and f^* denotes the grand maximal function of f, we have

$$c_\phi |\langle f, \phi \rangle| \leq f^*(x), \quad \text{all } x \in B.$$

Consequently, $|\langle f, \phi \rangle|^p \leq c_\phi' \|f\|_{H_w^p}^p$, and the conclusion follows at once. ∎

Lemma 3.. Suppose that $f \in \hat{\mathcal{D}}_0$, and let f^* denote the grand maximal function of f. Then, given an integer $N \geq 1$, there is a sequence $\{\beta_k\}$ of functions that satisfies the following properties:
  (i) $\beta_k = \sum_{i=1}^\infty \beta_{k,i}$, where each $\beta_{k,i} \in \mathcal{S}(R^n)$.
  (ii) $\operatorname{supp} \beta_{k,i} \subseteq B(x_i^k, c_1 r_i^k)$, where $\{B(x_i^k, r_i^k)\}$ is a collection of balls contained in the set $\{f^* \geq 2^k\}$, such that no point $x \in R^n$ belongs to more than finitely many balls, and c_1 is a fixed constant.
  (iii) $|\beta_{k,i}| \leq c_2$.
  (iv) All moments of $\beta_{k,i}$ up to order N vanish.
  (v) $f = \sum_{k=-\infty}^\infty \sum_i 2^k \beta_{k,i}$, where the convergence is understood to take place unconditionally in the pointwise and distribution senses.

Proof. Let

$$\mathcal{O}_k = \{f^* > 2^k\}, \quad k = 0 \pm 1, \pm 2 \ldots,$$

for each k, $\mathcal{O}_k$ is a bounded open set, possibly empty. We fix k for the time being, and consider a Whitney-like partition of unity $\{\phi_{k,i}\}$ of $\mathcal{O}_k$.

We adopt the following notational convention: Since the index k is fixed, it will be omitted in the subscripts, and it will be denote by a superscript "prime" when it corresponds to the value $k+1$. Thus, for instance, we write $\phi_{k,i} = \phi_i$ and $\phi_{k+1,j} = \phi_j'$.

The following properties of $\{\phi_i\}$ are useful to us:
  (i) $0 \leq \phi_i \leq 1$, $\operatorname{supp} \phi_i \subseteq B_i = B(x_i, r_i)$.
  (ii) $\sum_i \phi_i = \chi_{\mathcal{O}_k}$.
  (iii) $m_i = \int_{R^n} \phi_i(x)\, dx \sim r_i^n$.

Let $\mathcal{V}_{k,i} = \mathcal{V}_i$ denote the Hilbert space of polynomials p of degree less than or equal to N normed by

$$\|p_i\| = \left(\frac{1}{m_j} \int_{R^n} |p_i(x)|^2 \phi_i(x)\, dx \right)^{1/2};$$

the polynomials $\{(x - x_i)^\alpha\}_{|\alpha| \leq N}$ form a basis for $\mathcal{V}_i$. Let, then, $p_o, \ldots, p_L$ be the orthonormal basis of $\mathcal{V}_i$ obtained from this basis by the Gram-Schmidt orthogonalization process. It is readily seen that the coefficients of the p_j's, $0 \leq j \leq L$, are uniformly bounded. Further, let $P_i f$ denote the projection of f into $\mathcal{V}_i$, i.e.,

$$P_i f(x) = \sum_{j=0}^{L} \left(\frac{1}{m_i} \int_{R^n} f(y) P_j \phi_i(y) \, dy \right) p_j(x).$$

We claim that
$$|P_i f(x) \phi_i(x)| \leq c 2^k, \quad \text{all } i.$$

Indeed, since there is a constant c such that

$$\frac{c}{m_i} p_j(x) \phi_i(x) = \phi_j(y_i - x),$$

where $y_i \in B(x_i, a r_i) \cap O_k^c$, $a > 1$, and the ϕ_j's are the dilations of a function in the appropriate class, we see that

$$\left| \frac{1}{m_i} \int_{R^n} f(x) P_j \phi_i(x) \, dx \right| \leq c f^*(y_i) \leq c 2^k.$$

Similarly, if $P_{ij} f(x) = P_j'((f - P_j' f)\phi_i)(x)$, it readily follows that

(iv) $P_{ij} f \in \mathcal{V}_j'$, and $\sum_i (P_{ij} f) \phi_j' = 0$.

(v) $|P_{ij} f(x) \phi_j'(x)| \leq c 2^{k+1}$.

We decompose now f as follows: For each $k = 0, 1, \ldots$, put

$$f = \left(f \chi_{O_k^c} + \sum_i (P_i f) \phi_i \right) + \left(\sum_i (f - P_i f) \phi_i \right) = g_k + b_k,$$

say. Since $|f \chi_{O_k^c}| \leq 2^k$, it follows that $g_k \to 0$ as $k \to \infty$, actually $g_k = f$ for k sufficiently large. Thus, the telescoping series $f = \sum_{k=-\infty}^{\infty} (g_{k+1} - g_k)$ converges in the pointwise sense.

Next we show that we can write $g_{k+1} - g_k = 2^k \sum_i \beta_{k,i}$, with $\beta_{k,i}$ verifying conditions (ii)-(iv) in the statement of the lemma. Indeed, from the definition of the g_k's we get

$$g_{k+1} - g_k = \sum_i (f - P_i f)\phi_i - \sum_j (f - P_j' f)\phi_j'$$

$$= \sum_i \left(f - P_i f)\phi_i - \sum_j (f - P_j' f)\phi_j' \phi_i \right)$$

$$= \sum_i \left(f - P_i f)\phi_i - \sum_j ((f - P_j' f)\phi_i - \sum_j ((f - P_j' f)\phi_i - P_{ij} f)\phi_j' \right)$$

$$= \sum_i \beta_{k,i},$$

say. This explicit construction shows that the $\beta_{k,i}$'s have the desired properties; we only point out that at each point x there are only finitely many nonvanishing terms in all of the above sums. We also have

$$\sum_i |\beta_{k,i}| \leq \begin{cases} c2^k & \text{all } k\text{'s} \\ 0 & k \text{ large.} \end{cases}$$

Whence we conclude that $\sum_{i,k} \beta_{k,i}$ converges unconditionally in the pointwise and distribution sense to f. As the other properties are readily verified, we are done. ∎

We still need two preliminary results of independent interest.

Lemma 4. Suppose ν is a doubling weighted measure with respect to the Lebesgue measure on R^n, $\nu \in D_d$, with weight w. Then, for any sequence $\{B(x_k, r_k)\}$ of balls, any sequence $\{\lambda_k\}$ of positive numbers and $a \geq 1$, we have

$$\left\| \sum_k \lambda_k \chi_{B(x_k, ar_k)} \right\|_{L_w^p} \leq ca^\delta \left\| \sum_k \lambda_k \chi_{B(x_k, r_k)} \right\|_{L_w^p},$$

where

$$\delta = \begin{cases} d/p & 0 < p \leq 1 \\ d & 1 < p < \infty. \end{cases}$$

If in addition $w \in A_r$, the estimate holds with

$$\delta = \begin{cases} n + \left(\frac{1}{p} - \frac{1}{p}\frac{(p-1)}{(r-1)} \right)(d - n) & 1 < p < r \\ n & 1 < r \leq p. \end{cases}$$

Proof. The result is a direct consequence of Theorems 1-3 in Chapter IV; in fact the converse is also true. Let $\delta_{(x_k, r_k)}$ denote the Dirac delta measure concentrated at $(x_k, r_k) \in R_+^{n+1}$. Recall that

$$S_{a,1}(F, x) = \frac{1}{va^n} \int_{|x-y| \leq at} t^{-n} |F(y,t)| \, dy \frac{dt}{t}.$$

Now, replacing in this expression the measure $t^{-n}|F(y,t)|dy\frac{dt}{t}$ by the measure $\sum_k \lambda_k \delta_{(x_k, r_k)}$ we get

$$S_{a,1}\left(\sum_k \lambda_k \delta_{(x_k, r_k)}, x \right) = \frac{1}{va^n} \sum_k \lambda_k \chi_{B(x_k, ar_k)}(x);$$

similarly for $S_{1,1}$. That the conclusion of Theorems 1-3 in Chapter IV is also true for the "discrete" Lusin function $S_{a,1}(\sum_k \lambda_k \delta_{(x_k, r_k)})$ follows by approximating $\sum_k \lambda_k \delta_{B(x_k, r_k)}$ by continuous measures of the form $t^{-n}|F(y,t)|dy\frac{dt}{t}$, or else by a direct proof. ∎

Lemma 5. Suppose ν is a doubling weighted measure with respect to the Lebesgue measure in R^n, $\nu \in D_d$, with weight w. Further, if $q > \max(1,p)$, assume that $w \in RH_{(q/p)'}$. Then, for an arbitrary sequence $\{a_k\}$ of functions with support contained in $B(x_k, r_k)$ such that $\|a_k\|_{L^q} \leq \|\chi_{B(x_k,r_k)}\|_{L^q}$, and any sequence $\{\lambda_k\}$ of positive numbers, we have

$$\left\|\sum_k \lambda_k a_k\right\|_{L_w^p} \leq c \left\|\sum_k \lambda_k \chi_{B(x_k,r_k)}\right\|_{L_w^p}.$$

Proof. First we consider the case $0 < p \leq 1$. Observe that $(q/p)' = q/(q-p)$, and since $w^{(q/p)'} \in A_\infty$ and $(q-1)/(q-p) < 1$, by the remarks preceding Lemma 10 in Chapter I, we have

$$\left(\frac{1}{|B|} \int_B w(x)^{q/(q-p)}\, dx\right)^{1/q'} \leq c\left((w_B)^{q/(q-p)}\right)^{(q-1)/q}$$

$$= c\left(w^{(q-1)/(q-p)}\right)_B \leq c\frac{1}{|B|} \int_B w(x)^{(q-1)/(q-p)}\, dx.$$

Further, let $d\mu(x) = w(x)^{q/(q-p)}dx$, and $d\mu_1(x) = w(x)^{(q-1)/(q-p)}dx$. Put $B(x_k, r_k) = B_k$, and for $t > 0$ let

$$\mathcal{E} = \{y \in R^n : \sum_k \lambda_k \chi_{B_k}(y) > tw(y)^{1/(q-p)}\}$$

and

$$\mathcal{O} = \{y \in R^n : M_\mu \chi_{\mathcal{E}}(y) > 1/A\},$$

where A is a constant to be chosen shortly; by Theorem 3 in Chapter I, $\mu(\mathcal{O}) \leq cA\mu(\mathcal{E})$. From the above estimate we get

$$\int_{R^n} |a_k(x)|w(x)^{(q-1)/(q-p)}\, dx \leq \|a_k\|_{L^q} \left(\int_{B_k} w(x)^{q/(q-p)}\, dx\right)^{1/q'}$$

$$\leq \|\chi_{B_k}\|_{L^q}|B_k|^{1/q'} \left(\frac{1}{|B_k|} \int_{B_k} w(x)^{q/(q-p)}\, dx\right)^{1/q'}$$

$$\leq |B_k| \left(c\frac{1}{|B_k|} \int_{B_k} w(x)^{(q-1)/(q-p)}\, dx\right) = c\mu_1(B_k).$$

Now, if $y \in \mathcal{O}^c$ and B is a ball containing y, it follows that $\mu(\mathcal{E} \cap B)/\mu(B) \leq 1/A$. Whence, since $w(x)^{q/(q-p)}$ and $w(x)^{(q-1)/(q-p)}$ are A_∞ weights, if $\mathcal{O}^c \cap B_k \neq \emptyset$, there are constants $\gamma_1, \gamma_2 > 0$ such that

$$\mu_1(\mathcal{E} \cap B_k)/\mu_1(B_k) \leq c(|\mathcal{E} \cap B_k|/|B_k|)^{\gamma_1}$$
$$\leq c(\mu(B_k \cap \mathcal{E})/\mu(B_k))^{\gamma_2} \leq c(1/A)^{\gamma_2} < 1/2,$$

provided that A is large enough; this is our choice for A. This estimate in particular implies that if $\mathcal{O}^c \cap B_k \neq \emptyset$, then $\mu_1(B_k) \leq 2\mu_1(\mathcal{E}^c \cap B_k)$.

Next we claim that

$$\int_{\mathcal{O}^c} \left| \sum_k \lambda_k a_k(x) \right| d\mu_1(x) \leq c \int_{\mathcal{E}^c} \left(\sum_k \lambda_k \chi_{B_k}(x) \right) d\mu_1(x).$$

Indeed, by the above observations, the left hand-side above is dominated by

$$\sum_k \lambda_k \int_{B_k} |a_k(x)| \, d\mu_1(x) \leq c \sum_k \lambda_k \mu_1(B_k)$$

$$\leq c \sum_k \lambda_k \mu_1(B_k \cap \mathcal{E}^c) = c \int_{\mathcal{E}^c} \sum_k \lambda_k \chi_{B_k}(x) \, d\mu_1(x),$$

as asserted. Let

$$f = \sum_k \lambda_k a_k(x) w(x)^{-1/(q-p)}, \quad g = \sum_k \lambda_k \chi_{B_k}(x) w(x)^{-1/(q-p)}.$$

Since $d\mu(x) = w(x)^{-1/(q-p)} d\mu_1(x)$, the above estimate implies

$$\mu(\{|f| > t\}) \leq \mu(\mathcal{O}) + \frac{1}{t} \int_{\mathcal{O}^c} f(x) \, d\mu(x)$$

$$\leq c\mu(\mathcal{E}) + \frac{1}{t} \int_{\mathcal{E}^c} g(x) \, d\mu(x)$$

$$\leq c\mu(\{g > t\}) + c\frac{1}{t} \int_{\{g \leq t\}} g(x) \, d\mu(x).$$

Whence, since $\|\phi\|_{L_w^p} = \|\phi/w^{1/(q-p)}\|_{L^p(\mu)}$, multiplying through by pt^{p-1} and integrating it follows that

$$\left\| \sum_k \lambda_k a_k \right\|_{L_w^p} \leq c \left\| \sum_k \lambda_k \chi_{B_k} \right\|_{L_w^p} \leq c \left\| \sum_k \lambda_k^s \chi_{B_k} \right\|_{L_w^{p/s}}, \quad 0 < s < 1.$$

This completes the proof of the case $0 < p \leq 1$. As for the case $1 < p < \infty$, first pick p_0 so that $q' < p_0' < p'$ and $w \in RH_{(r/p_0)}$, where $1/q + 1/p_0' + 1/r = 1$. Then by Hölder's inequality, and with $\|g\|_{L_w^{p'}} \leq 1$, we have

$$\left| \int_{R^n} a_k(y) g(y) w(y) \, dy \right| \leq \|a_k\|_{L^q} \|\chi_{B_k} g\|_{L_w^{p_0'}} \|\chi_{B_k}\|_{L^{r/p_0}}^{1/p_0}$$

$$\leq |B_k|^{1/q} \nu(B_k)^{1/p_0'} \left(\frac{1}{\nu(B_k)} \int_{B_k} g(x)^{p_0'} w(x) \, dx \right)^{1/p_0'}$$

$$\times |B_k|^{1/r} \left(\frac{1}{|B_k|} \int_{B_k} w(x)^{r/p_0} \, dx \right)^{1/r}$$

$$\leq |B_k|^{1/p_0} \nu(B_k)^{1/p_0'} \inf_{x \in B_k} M_\nu(g^{p_0'})(x)^{1/p_0'} \left(\frac{1}{|B_k|} \int_{B_k} w(x) \, dx \right)^{1/p_0}$$

$$\leq c\nu(B_k) \inf_{x \in B_k} M_\nu(g^{p_0'})(x)^{1/p_0'} \leq c \int_{B_k} M_\nu(g^{p_0'})(x)^{1/p_0'} w(x) \, dx.$$

Thus,

$$\left| \int_{R^n} \left(\sum_k \lambda_k a_k(x) \right) g(x) w(x) \, dx \right| \leq c \sum_k \lambda_k \int_{B_k} M_\nu(g^{p_0'})(x)^{1/p_0'} w(x) \, dx$$

$$\leq \int_{R^n} \left(\sum_k \lambda_k \chi_{B_k}(x) \right) M_\nu(g^{p_0'})(x)^{1/p_0'} w(x) \, dx$$

$$\leq c \left\| \sum_k \lambda_k \chi_{B_k}(x) \right\|_{L_w^p} \left\| M_\nu(g^{p_0'})^{1/p_0'} \right\|_{L_w^{p'}}$$

$$\leq c \left\| \sum_k \lambda_k \chi_{B_k} \right\|_{L_w^p},$$

and the proof is complete. ∎

Proof of Theorem 1. When $f \in \hat{\mathcal{D}}_0$ the first half of the theorem has essentially been established in Lemma 3; before we do the decomposition of an arbitrary f in $H_w^p(R^n)$ we prove the second half of the theorem. Thus, we assume that

$$\left\| \sum_k \lambda_k^s \chi_{B_k} \right\|_{L_w^{p/s}} < \infty, \quad 0 < s \leq 1.$$

Since in this case also

$$\left\| \sum_k \lambda_k \chi_{B_k} \right\|_{L_w^p} \leq \left\| \sum_k \lambda_k^s \chi_{B_k} \right\|_{L_w^{p/s}} < \infty,$$

we may as well take $s = 1$. We want thus to show that if the b_k's are (N, q) atoms with support contained in B_k, and $N \geq N(q, w)$, $1 < q \leq \infty$, and if in addition $w \in RH_{(q/p)'}$ when $1 < q < \infty$, then

$$\left\| \sum_k \lambda_k b_k \right\|_{H_w^p} \leq c \left\| \sum_k \lambda_k \chi_{B_k} \right\|_{L_w^p}.$$

If b is then a (q, N) atom with support contained in $B(z, r)$, we estimate the radial maximal function of $F(x, t) = b * \phi_t(x)$, where ϕ is a Schwartz function with support contained in $B = B(0, 1)$ and integral 1. First, if $x \in B(z, 2r)$, we invoke the fact that $M_0(F, x)$ is dominated by the Hardy-Littlewood maximal function of b. Thus, by Theorem 3 in Chapter I,

$$\|\chi_{B(z,2r)} M_0(F)\|_{L^q} \leq c\|b\|_{L^q} \leq c\|\chi_{B(z,r)}\|_{L^q} \leq c\|\chi_{B(z,2r)}\|_{L^q},$$

and $\chi_{B(z,2r)}(x) M_0(F, x)$ is a multiple of a $(q, -1)$ atom. Now, if $x \notin B(z, 2r)$, observe that $F(x, t) = 0$ if $t < |x|/2$. Indeed, if $y \in B(0, r)$, then $|x - y| \geq |x|/2$, and at least one of the factors in the integrand of $\int_{R^n} b(y)\phi_t(x - y)\, dy$ vanishes for any given y. Also, for $t \geq |x|/2$, we let p_ϕ denote the Taylor expansion of ϕ of degree N about the point x/t. Then,

$$|F(x, t)| = t^{-n} \left| \int_{R^n} (\phi((x - y)/t) - p_\phi((-y)/t)) b(y)\, dy \right|$$

$$\leq ct^{-n} \int_{R^n} (|y|/t)^{N+1} |b(y)|\, dy.$$

Since $\|b\|_{L^q} \leq cr^{n/q}$, by Hölder's inequality we see that in this case

$$|F(x, t)| \leq c(|x|/r)^{-(n+N+1)},$$

and consequently, $M_0(F, x) \leq c(|x|/r)^{-(n+N+1)}$. Therefore, with $c_k \leq c$, it follows that $M_0(F, x)$ is dominated by

$$\chi_{B(z,2r)}(x) M_0(F, x) + \sum_{k=1}^{\infty} c_k(|x|/t)^{-(n+N+1)} \left(\chi_{B(z,2^{k+1}r)}(x) - \chi_{B(z,2^k,r)}(x) \right)$$

$$\leq ca_0(x) + c\sum_{k=1}^{\infty} 2^{-k(n+N+1)} a_k(x),$$

say, where the a_k's are $(q, -1)$ atoms with support contained in $B(z, 2^{k+1}r)$. Further, since M_0 is subadditive, we can apply the above estimate to finite linear combinations of atoms. First, in the case $w \in A_p$, $M_0((\sum_{j=1}^{L} \lambda_j b_j) * \phi_t, x)$ is dominated by the Hardy-Littlewood maximal function of $\sum_{j=1}^{L} \lambda_j b_j$, which is bounded, in the $L_w^p(R^n)$-norm, by $\|\sum_{j=1}^{L} \lambda_j b_j\|_{L_w^p}$. Thus, by Theorem 12 in Chapter IV and Lemma 5, we have

$$\left\| \sum_{j=1}^{L} \lambda_j b_j \right\|_{H_w^p} \leq c \left\| \sum_{j=1}^{L} \lambda_j \chi_{B(z_j, r_j)} \right\|_{L_w^p}.$$

In the other cases, by the above observations, we get

$$M_0\left(\left(\sum_{j=1}^{L} \lambda_j b_j \right) * \phi_t, x \right) \leq c \sum_{j=1}^{L} \sum_k \lambda_j 2^{-k(n+N+1)} a_{j,k}(x),$$

where the $a_{j,k}$'s are $(q,-1)$ atoms with support contained in the ball $B_{j,k} = B(z_j, 2^{k+1}r_j)$. Thus, by Lemma 4, and if $q < \infty$ also by Lemma 5, with $p_1 = \min(1,p)$, we have

$$\left\| \sum_{j=1}^{L} \lambda_j b_j \right\|_{H_w^p}^{p_1} \leq c \left\| \sum_{j=1}^{L} \sum_k \lambda_j 2^{-k(n+N+1)} a_{j,k} \right\|_{L_w^p}^{p_1}$$

$$\leq c \sum_k 2^{-k(n+N+1)p_1} \left\| \sum_{j=1}^{L} \lambda_j \chi_{B_{j,k}} \right\|_{L_w^p}^{p_1}$$

$$\leq c \left(\sum_k 2^{-k(n+N+1)p_1} 2^{k\delta} \right) \left\| \sum_{j=1}^{L} \lambda_j B_{(z_j,r_j)} \right\|_{L_w^p}^{p_1}.$$

Now, if $\delta < (n+N+1)p_1$, the above sum converges, and again we get

$$\left\| \sum_{j=1}^{L} \lambda_j b_j \right\|_{H_w^p} \leq c \left\| \sum_{j=1}^{L} \lambda_j B_{(z_j,r_j)} \right\|_{L_w^p}.$$

If the linear combination $\sum_{j=1}^{\infty} \lambda_j \chi_{B(z_j,r_j)}$ is now infinite, then the partial sums $\{\sum_{j=1}^{M} \lambda_j \chi_{B(z_j,r_j)}\}_{M=1}^{\infty}$ form a Cauchy sequence in $L_w^p(R^n)$. From the above estimates we get that $\{\sum_{j=1}^{M} \lambda_j b_j\}_{M=1}^{\infty}$ is also a Cauchy sequence in $H_w^p(R^n)$ and consequently, it converges to some distribution $f \in H_w^p(R^n)$, say, in the $H_w^p(R^n)$-norm. By Lemma 2 the convergence is also in the sense of distributions, and we have

$$\|f\|_{H_w^p} \leq c \left\| \sum_j \lambda_j \chi_{B(z_j,r_j)} \right\|_{L_w^p}.$$

This completes the proof of the second part of Theorem 1.

To complete the proof of the first half, assume first that $f \in \hat{D}_0$ and decompose it as in Lemma 3 with $N > N(p,w)$. We thus get a sum $\sum_{j,k} \lambda_{j,k} \beta_j^k$ with $\lambda_{j,k} = 2^k$ and β_j^k (∞, N) atoms with support contained in balls B_j^k, where $\bigcup_j B_j^k \subseteq \mathcal{U}_k = \{f^* > c2^k\}$, c an appropriate constant, and so that, uniformly in k, the collection $\{B_j^k\}$ has the finite overlapping property. It then follows that for $0 < s < \infty$,

$$\sum_j \lambda_{j,k}^s \chi_{B_j^k}(x) \leq c 2^{ks} \chi_{\mathcal{U}_k}(x),$$

and

$$\sum_{j,k} \lambda_{j,k}^s \chi_{B_j^k}(x) \leq c \sum_k 2^{ks} \chi_{\mathcal{U}_k}(x) \leq c f^*(x)^s.$$

This gives the estimate

$$\left\| \sum_{j,k} \lambda_{j,k}^s \chi_{B_j^k} \right\|_{L_w^{p/s}} \leq c_s \|f^*\|_{L_w^p} \leq c_s \|f\|_{H_w^p} \quad 0 < s < \infty.$$

Now we show that $\sum_{j,k} \lambda_{j,k} \beta_j^k(x)$ converges to f in the $H_w^p(R^n)$-norm. Observe that by Lemma 3 the sum converges to f in the sense of distributions. Furthermore, by the second half of the theorem, which we just proved, the sum converges in the $H_w^p(R^n)$-norm and in the sense of distributions to $g \in H_w^p(R^n)$; since it also converges to f in the sense of distributions, it follows that $g = f$.

Finally, for an arbitrary f in $H_w^p(R^n)$, we choose a sequence $\{f_m\} \subseteq \hat{\mathcal{D}}_0$ such that it tends to f in the $H_w^p(R^n)$-norm as $m \to \infty$, $\|f_1\|_{H_w^p} \leq 2\|f\|_{H_w^p}$, and $\|f_m - f_{m-1}\|_{H_w^p} \leq 2^{-m}\|f\|_{H_w^p}$ for $m \geq 2$. Putting $g_1 = f_1$, $g_m = f_m - f_{m-1}$ for $m \geq 2$, it follows that $f = \sum_{m=1}^{\infty} g_m$, with convergence in the $H_w^p(R^n)$-norm. Let $g_m = \sum_k \lambda_{k,m} b_{k,m}$ denote the atomic decomposition given in Lemma 3 of $g_m \in \hat{\mathcal{D}}_0$. Observe that $\sum_{k,m} \lambda_{k,m} b_{k,m}$ converges unconditionally to f and that if the balls $B_{k,m}$ contain the supports of the atoms and $p_1 = \min(1,p)$, then

$$\left\| \sum_{k,m} \lambda_{k,m}^s \chi_{B_{k,m}} \right\|_{L_w^{p/s}}^{p_1} \leq \sum_m \left\| \sum_k \lambda_{k,m}^s \chi_{B_{k,m}} \right\|_{L_w^{p/s}}^{p_1}$$

$$\leq c \sum_m \|g_m\|_{H_w^p}^{p_1} \leq c \left(\sum_m 2^{-mp_1} \right) \|f\|_{H_w^p}.$$

This completes the proof. ∎

Sources and Remarks. The atomic decomposition of distributions in $H^p(R)$ obtained by R. Coifman [1974] was extended by R. Latter [1977] to the spaces $H^p(R^n)$, $0 < p \leq 1$, and by J. García-Cuerva [1979] to some weighted $H_w^p(R^n)$ spaces, again with $0 < p \leq 1$ and with w in an appropriate A_q class. The proof of Lemma 3 is due to Latter.

CHAPTER

IX

The Basic Inequality

An important role in the theory of weighted Hardy spaces is played by what we call here the basic inequality. This is roughly a principle which allows, among other applications, for the use of maximal functions to describe the dual to the Hardy spaces and to control the various operators that act naturally on these spaces. We present in Theorem 2 and 3 below two versions of the inequality. Theorem 2 holds for doubling weights and the proof uses the atomic decomposition. On the other hand, Theorem 3, which is a particular case of Theorem 2, uses only elementary properties of the heat kernel, but it is only valid for A_∞ weights.

We begin by discussing a third, and elementary, version of the inequality; this version is, however, sufficient for many applications.

Proposition 1. Let ν be a doubling weighted measure with respect to the Lebesgue measure on R^n, $\nu \in D_d$, with weight w, and suppose that $f \in \hat{\mathcal{D}}_0$ and $g \in L^1_{\mathrm{loc}}(R^n)$. Then

$$\left| \int_{R^n} f(x)g(x)\, dx \right| \le c \int_{R^n} f^*(x) M^{\sharp,\mathcal{P}}_{1,\nu} g(x)\, w(x)\, dx \,,$$

with c independent of f, g and $\mathcal{P}$ an appropriate class of polynomials of degree less that or equal to $N = N(d)$.

Proof. Let $f = \sum_k \sum_i 2^k \beta_{k,i}$ be the atomic decomposition of f given in Lemma 3 of Chapter VIII; this decomposition is built from the grand maximal function f^* corresponding to a Schwartz function with nonvanishing integral and it has the property that there are balls $B_{k,i} \supseteq \mathrm{supp}\ \beta_{k,i}$ with the finite overlapping property, such that $\sum_{k,i} 2^k \chi_{B_{k,i}}(x) \le cf^*(x)$. As in Chapter III, given a ball B, let $p_B(g)$ be a function in $\mathcal{P}$ such that

$$A_1^P(g, B) = \frac{1}{\nu(B)} \int_B |g(x) - p_B(g)(x)| \, dx \le \inf_{x \in B} M_{1,\nu}^{\sharp,P} g(x).$$

Then,

$$I = \int_{R^n} f(x)g(x) \, dx = \sum_k \sum_i 2^k \int_{B_{k,i}} \beta_{k,i}(x)g(x) \, dx$$

$$= \sum_k \sum_i 2^k \int_{B_{k,i}} \beta_{k,i}(x) \left(g(x) - p_{B_{k,i}}(g)(x) \right) dx. \tag{1}$$

At this point we should really say a word about the convergence of the integral defining I. For $f \in \hat{\mathcal{D}}_0$ we have $f^*(x) \le c(1 + |x|)^{-N}$ for any $N > 0$, and if $M_{1,\nu}^{\sharp,P} g(x) < \infty$ for some x, then we can control how fast the averages of g will grow on large balls about the origin. From this we see that $\int_{R^n} f^*(x)|g(x)|w(x) \, dx < \infty$, and the convergence of the integral defining I follows readily.

Returning to the proof, then, observe that $|I|$ is bounded by

$$c\sum_k \sum_i 2^k \nu(B_{k,i}) A_1^P(g, B_{k,i}) \le c\sum_k \sum_i 2^k \nu(B_{k,i}) \inf_{x \in B_{k,i}} M_{1,\nu}^{\sharp,P} g(x)$$

$$\le c\sum_k \sum_i 2^k \int_{B_{k,i}} M_{1,\nu}^{\sharp,P} g(x)w(x) \, dx$$

$$= c \int_{R^n} \sum_{k,i} 2^k \chi_{B_{k,i}}(x) M_{1,\nu}^{\sharp,P} g(x)w(x) \, dx$$

$$\le c \int_{R^n} f^*(x) M_{1,\nu}^{\sharp,P} g(x)w(x) \, dx. \quad \blacksquare$$

Our next result improves on Proposition 1; to state it we need a notation. Given $A \ge 1$, let

$$M_{1,\nu,A}^{\sharp,P} g(x) = \sup_{B \supseteq \{x\}} \inf_{p \in \mathcal{P}} \frac{1}{\nu(B)^A} \int_B |g(x) - p(x)| \, dx.$$

We then have,

Theorem 2. Let ν be a doubling weighted measure with respect to the Lebesgue measure on R^n, $\nu \in D_d$, $d\nu(x) = w(x)dx$, and suppose that $f \in \hat{\mathcal{D}}_0$, and $g \in L_{\text{loc}}^1(R^n)$. Then, given a Schwartz function ϕ with nonvanishing integral and a constant $A \ge 1$, there is a constant c independent of f, g such that

$$\left| \int_{R^n} f(x)g(x) \, dx \right| \le c \left(\int_{R^n} \left(M_1(f * \phi_t, x) M_{1,\nu,A}^{\sharp,P} g(x) \right)^{1/A} w(x) \, dx \right)^A.$$

124

Proof. Let $F(x,t) = f * \phi_t(x)$, and pick λ so that $\lambda > dA$. Further recall that, properly normalized, $f^*(x) \leq cN_\lambda(f,x)$.

Let $\mathcal{F}_j$ denote the collection of those balls $B(y,t)$ such that

$$2^j < |F(y,t)| \leq 2^{j+1}, \quad j = 0, \pm1, \pm2, \ldots$$

Since $f \in \hat{\mathcal{D}}_0$, $|F(y,t)| \leq \|f\|_{L^1} t^{-n} \|\phi\|_{L^\infty}$. Hence, for each y, $|F(y,t)| \to 0$ as $t \to \infty$ and for each fixed j, all balls in $\mathcal{F}_j$ have bounded radius. We are, then, under the conditions of the covering lemma in Chapter I. If a denotes the constant in the covering lemma, and as usual $aB(y,t) = B(y,at)$, let $\mathcal{F}'_j = \{B'\}$ be a countable pairwise disjoint collection of balls in $\mathcal{F}_j$ with the property that to each ball $B \in \mathcal{F}_j$ there corresponds $B' \in \mathcal{F}'_j$ such that $B \subseteq aB'$. Now put $c(j) = a2^{j/\lambda}$; we claim that

$$\{N_\lambda(F) > 2^k\} \subseteq \bigcup_{h \geq k} \bigcup_{B \in \mathcal{F}_h} 2^{(h-k)/\lambda} B \subseteq \bigcup_{h \geq k} \bigcup_{B' \in \mathcal{F}'_h} c(h-k)B' = \mathcal{U}_k,$$

say. Indeed, suppose that $N_\lambda(F,x) > 2^k$, and let $h_0 = \max\{h : M_{2^{h/\lambda}}(F,x) > 2^{k+h}\}$, cf. the proof of Theorem 11 in Chapter IV. Then there is $(y,t) \in R_+^{n+1}$ such that $|x - y| < 2^{h_0/\lambda} t$ and $2^{k+h_0} < |F(y,t)| \leq 2^{k+h_0+1}$. Thus $x \in 2^{h_0/\lambda} B$ with $B \in \mathcal{F}_{k+h_0}$ and, as asserted, $\{N_\lambda(F) > 2^k\} \subseteq \bigcup_{h \geq 0} \bigcup_{B \in \mathcal{F}_h} 2^{h/\lambda} B \subseteq \mathcal{U}_k$.

Next consider the atomic decomposition $f = \sum_k \sum_i 2^k \beta_{k,i}$ described in Proposition 1; clearly $B_{k,i} \subseteq \{N_\lambda(F) > 2^k\}$ and $\bigcup_i B_{k,i} \subseteq \mathcal{U}_k$. Having fixed k for the time being, we separate the $B_{k,i}$'s into two disjoint families as follows:

1. Those balls for which there are $h \geq k$ and $B' \in \mathcal{F}'_h$ such that $B_{k,i} \subseteq 2c(h-k)B'$; call them B_{k,i_1}.
2. $B_{k,i} \not\subseteq 2c(h-k)B'$ for any $B' \in \mathcal{F}'_h$, $h \geq k$; we call these balls B_{k,i_2}.

A word about the second family. Since each $B_{k,i_2} \subseteq \mathcal{U}_k$,

$$B_{k,i_2} \cap c(h-k)B' \neq \emptyset, \quad \text{some } h \geq k \quad \text{and} \quad B' \in \mathcal{F}'_h, \tag{2}$$

in this case we have

$$\bigcup_{B' \in \mathcal{F}'_h, (2)\text{holds}} B' \subseteq 5B_{k,i_2},$$

and

$$B_{k,i_2} \subseteq \bigcup_{h \geq k} \bigcup_{B' \in \mathcal{F}'_h, (2)\text{holds}} B'.$$

Now, in the notation of Proposition 1, we estimate $|I|$ by

$$c\sum_k 2^k \sum_{i_1} \int_{B_{k,i_1}} |g(x) - p_{B_{k,i_1}}(g)(x)| \, dx$$

$$+ c\sum_k 2^k \sum_{i_2} \int_{B_{k,i_2}} |g(x) - p_{B_{k,i_2}}(g)(x)| \, dx = \sum_k 2^k I_1(k) + c\sum_k 2^k I_2(k),$$

say. We estimate $I_1(k)$ first. Fix, in addition to k, $h \geq k$ and a ball $B' \in \mathcal{F}'_h$ and consider those balls B_{k,i_1} such that

$$B_{k,i_1} \subseteq 2c(h-k)B'. \tag{3}$$

Then,

$$
\begin{aligned}
I_{1,h,B'}(k) &= \sum_{i_1, (3)\text{holds}} \int_{B_{k,i_1}} |g(x) - p_{B_{k,i_1}}(g)(x)| \, dx \\
&\leq \int_{2c(h-k)B'} |g(x) - p_{2c(h-k)B'}(g)(x)| \, dx \\
&\leq \nu(2c(h-k)B')^A \inf_{x \in B'} M^{\sharp,\mathcal{P}}_{1,\nu^A} g(x) \leq c 2^{(h-k)dA/\lambda} \nu(B')^A \inf_{x \in B'} M^{\sharp,\mathcal{P}}_{1,\nu^A} g(x) \\
&\leq c 2^{(h-k)dA/\lambda} \left(\int_{B'} M^{\sharp,\mathcal{P}}_{1,\nu^A} g(x) \, dx \right)^A.
\end{aligned}
$$

Clearly $I_1(k) \leq \sum_{h \geq k} \sum_{B' \in \mathcal{F}'_h} I_{1,h,B'}(k)$. Thus, the sum $\sum_k 2^k I_1(k)$ is dominated by

$$
\begin{aligned}
\sum_k 2^k &\sum_{h \geq k} 2^{(h-k)dA/\lambda} \sum_{B' \in \mathcal{F}'_h} \left(\int_{B'} M^{\sharp,\mathcal{P}}_{1,\nu^A} g(x)^{1/A} w(x) \, dx \right)^A \\
&= c \sum_{h=-\infty}^{\infty} \sum_{B' \in \mathcal{F}'_h} \left(\int_{B'} M^{\sharp,\mathcal{P}}_{1,\nu^A} g(x)^{1/A} w(x) \, dx \right)^A 2^h \sum_{k \leq h} 2^{(k-h)(1-dA/\lambda)} \\
&\leq c \sum_{h=-\infty}^{\infty} 2^h \sum_{B' \in \mathcal{F}'_h} \left(\int_{B'} M^{\sharp,\mathcal{P}}_{1,\nu^A} g(x)^{1/A} w(x) \, dx \right)^A \\
&\leq c \left(\sum_{h=-\infty}^{\infty} \int_{\mathcal{F}'_h} 2^{h/A} M^{\sharp,\mathcal{P}}_{1,\nu^A} g(x)^{1/A} w(x) \, dx \right)^A \\
&\leq c \left(\int_{R^n} \sum_{h=-\infty}^{\infty} 2^{h/A} \chi_{\mathcal{F}'_h}(x) M^{\sharp,\mathcal{P}}_{1,\nu^A} g(x)^{1/A} w(x) \, dx \right)^A.
\end{aligned}
$$

Now, since $\mathcal{F}'_h \subseteq \{M_1(F) > 2^h\}$, the last integral above is dominated by

$$
c \left(\int_{R^n} \left(\sum_{h, 2^h \leq M(F,x)} 2^{h/A} \right) M^{\sharp,\mathcal{P}}_{1,\nu^A} g(x)^{1/A} w(x) \, dx \right)^A
$$

$$
\leq c \left(\int_{R^n} \left(M_1(F,x) M^{\sharp,\mathcal{P}}_{1,\nu^A} g(x) \right)^{1/A} w(x) \, dx \right),
$$

which is an estimate of the right order.

The proof for the B_{k,i_2}'s is similar. Indeed, for each fixed k we have

$$
\begin{aligned}
I(k, i_2) &= \int_{B_{k,i_2}} |g(x) - p_{B_{k,i_2}}(g)(x)| \, dx \\
&\leq c\nu(2B_{k,i_2})^A \inf_{x \in B_{k,i_2}} M_{1,\nu^A}^{\sharp,\mathcal{P}} g(x) \\
&\leq c\left(\sum_{h \geq k} \sum_{B' \in \mathcal{F}_h'} \nu(c(h-k)B') \inf_{x \in B'} M_{1,\nu^A}^{\sharp,\mathcal{P}} g(x)^{1/A} \right)^A,
\end{aligned}
$$

where the sum is extended over those $B' \in \mathcal{F}_h'$ in the definition of the case 2 for B_{k,i_2}. Whence, it follows that

$$
I_2(k) \leq c \left(\sum_{h \geq k} 2^{(h-k)d/\lambda} \sum_{B' \in \mathcal{F}_h'} \int_{B'} M_{1,\nu^A}^{\sharp,\mathcal{P}} g(x)^{1/A} w(x) \, dx \right)^A,
$$

and consequently, the sum $\sum_k 2^k I_2(k)$ does not exceed

$$
\begin{aligned}
c \sum_{k=-\infty}^{\infty} & \left(\sum_{h \geq k} 2^{(h-k)d/\lambda} 2^{(k-h)/A} \sum_{B' \in \mathcal{F}_h'} \int_{B'} 2^{h/A} M_{1,\nu^A}^{\sharp,\mathcal{P}} g(x)^{1/A} w(x) \, dx \right)^A \\
&\leq c \left(\int_{R^n} \sum_{h=-\infty}^{\infty} 2^{h/A} \chi_{\mathcal{F}_h'}(x) M_{1,\nu^A}^{\sharp,\mathcal{P}} g(x)^{1/A} w(x) \, dx \right)^A \\
&\leq c \left(\int_{R^n} \left(M(F,x) M_{1,\nu^A}^{\sharp,\mathcal{P}} g(x) \right)^{1/A} w(x) \, dx \right)^A.
\end{aligned}
$$

Since this estimate is also of the right order, we are done. ∎

We state now and prove Theorem 3.

Theorem 3. Let ν be a weighted measure with respect to the Lebesgue measure on R^n with weight $w \in A_\infty$, and $\phi(x) = e^{-\pi|x|^2}$. If $f \in \hat{\mathcal{D}}_0$ and g is locally integrable, and if $F(x,t) = f * \phi_t(x)$, then there is a constant c independent of f, g such that

$$
\left| \int_{R^n} f(x)g(x) \, dx \right| \leq c \int_{R^n} M(F,x) M_{1,\nu}^{\sharp} g(x) w(x) \, dx.
$$

Proof. We achieve the proof through a number of steps of independent interest which we label lemmas.

Lemma 4. Let $G(y,t) = t|\nabla F(y,t)|$ and fix a ball $B = B(x,h)$. Given $0 < s_0 < 1$, there are constants $a, \bar{s}_0$, such that for all $\lambda > 0$,

$$|\{y \in B : M_2^{0,2h}(F,y) > a\lambda\}| < \bar{s}_0|B| \quad \text{implies}$$
$$|\{y \in B : S_1^{0,h}(G,y) > \lambda\}| < s_0|B|.$$

Proof. It follows from Lemma 11 in Chapter VI. By the mean value inequality, $M_2^{0,2h}(N,x)$ in that lemma is dominated by a multiple of $M_2^{0,2h}(F,x)$, and the conclusion follows from this at once. ∎

Lemma 5. Let the functions F, G and the ball B be as in Lemma 4. Given $w \in A_\infty$, let $\operatorname{av}_B w$ denote the average of w over B. Then, given $0 < s_1 < 1$, there are constants $a_1, \bar{s}_1$ so that for all $\mu > 0$,

$$|\{y \in B : M_2^{0,2h}(F,y) > a_1 \mu \operatorname{av}_B w\}| < \bar{s}_1|B| \quad \text{implies}$$
$$|\{y \in B : S_1^{0,h}(G,y) > \mu w(y)\}| < s_1|B|.$$

Proof. Let $1 < p < \infty$ be such that $w \in A_p$, and consider the set $\mathcal{E} = \{y \in B : w(y) < b \operatorname{av}_B w\}$; here b is a constant to be chosen. Then there is a constant c such that $|\mathcal{E}| \le c^{p/(p-1)} b^{1/(p-1)}|B|$, and consequently, we can choose b so that $|\mathcal{E}| < s_1|B|/2$. Also observe that with this choice of b,

$$|\{y \in B, : S_1^{0,h}(G,y) > b\mu \operatorname{av}_B w\}| < s_1|B|/2 \quad \text{implies}$$
$$|\{y \in B : S_1^{0,h}(G,y) > \mu w(y)\}| < s_1|B|.$$

Indeed, since

$$|\{y \in B : S_1^{0,h}(G,y) > \mu w(y)\}|$$
$$\le |\{y \in B \cap \mathcal{E}^c : S_1^{0,h}(G,y) > \mu w(y)\}| + |\mathcal{E}|$$
$$\le |\{y \in B : S_1^{0,h}(G,y) > b\mu \operatorname{av}_B w\}| + s_1|b|/2 < s_1|B|.$$

Now we just apply Lemma 4 with $s_0 = s_1$ and $\lambda = \mu \operatorname{av}_B w$, and combine this with the above implication to obtain the desired conclusion. ∎

Lemma 6. In the setting of Lemma 5, suppose that $\nu(2\tilde{B}) \le A\nu(\tilde{B})$ for all balls $\tilde{B}$. Then, for any locally integrable function g and any ball $B = B(x,h)$ we have

$$\int_{R^n} |g(y) - p_B(g)(y)|\phi_h(x-y)\,dy \le c \operatorname{av}_B w \inf_{y \in B} M_{1,\nu}^\sharp g(y).$$

Proof. Let $\eta \in C_0^\infty(R^n)$, $\text{supp } \eta \subseteq \{|y| \leq 1\}$, be such that $\eta_t * p(y) = p(y)$ for all polynomials p of degree less than or equal to $N = N(d)$, and all $t > 0$. Then for $y \in 2^{k-2}B$, and as a consequence for $y \in 2^{k+1}B$, there is a constant c such that

$$|p_{2^k B}(g)(y) - p_{2^{k-1}B}(g)(y)| \leq c(A/2^n)^k M_{1,\nu}^{\sharp}g(y) \, \text{av}_B w.$$

Indeed, we have

$$|p_{2^k B}(g)(y) - p_{2^{k-1}B}(g)(y)|$$

$$= \left| \frac{1}{(2^{k-2}h)^n} \int_{R^n} \eta((2^{k-2}h)^{-n}(y-z))(p_{2^k B}(g)(z) - p_{2^{k-1}B}(g)(z)) \, dz \right|$$

$$\leq \frac{c}{(2^{k-1}h)^n} \int_{2^{k-1}B} |p_{2^{k-1}B}(g)(z) - g(y-z)| \, dz$$

$$+ \frac{c}{(2^{k-1}h)^n} \int_{2^k B} |p_{2^k B}(g)(z) - g(y-z)|$$

$$\leq \frac{c}{(2^{k-1}h)^n} \left(\nu(2^{k-1}B) + \nu(2^k B) \right) M_{1,\nu}^{\sharp}g(y) \leq c(A/2^n)^k M_{1,\nu}^{\sharp}g(y) \, \text{av}_B w.$$

Whence, if $z \in B$, it follows that

$$\int_{R^n} |g(y) - p_B(g)(y)| \phi_h(x-y) \, dy$$

$$\leq \frac{1}{h^n} \int_B |g(y) - p_B(g)(y)| \phi((x-y)/h) \, dy$$

$$+ \sum_{k=1}^{\infty} \frac{1}{h^n} \int_{2^k B \setminus 2^{k-1}B} |g(y) - p_B(g)(y)| \phi((x-y)/h) \, dy$$

$$\leq M_{1,\nu}^{\sharp}g(z) \, \text{av}_B w + \sum_{k=1}^{\infty} \phi(2^{k-1}) \frac{1}{h^n} \int_{2^k B} |g(y) - p_{2^k B}(g)(y)| \, dy$$

$$+ \sum_{k=1}^{\infty} \phi(2^{k-1}) \frac{1}{h^n} |2^k B| \sup_{y \in 2^k B} |p_{2^j B}(g)(y) - p_{2^j B}(g)(y)|.$$

It thus only remains to bound the two sums above. The first sum does not exceed

$$\sum_{k=1}^{\infty} \phi(2^{k-1}) \frac{1}{h^n} \nu(2^k B) M_{1,\nu}^{\sharp}(g)(z)$$

$$\leq c M_{1,\nu}^{\sharp}(g)(z) \, \text{av}_B w \sum_{k=1}^{\infty} \phi(2^{k-1})(A/2^n)^k 2^{nk} \leq c M_{1,\nu}^{\sharp}(g)(y) \, \text{av}_B w.$$

As for the other sum, it is bounded by

$$c \sum_{k=1}^{\infty} \phi(2^{k-1}) \frac{1}{h^n} (2^k h)^n \sum_{j=1}^{k} (A/2^n)^j M_{1,\nu}^{\sharp}(g)(z) \, \text{av}_B w$$

$$\leq c \left(\sum_{k=1}^{\infty} \phi(2^{k-1}) 2^{nk} A^k \right) M_{1,\nu}^{\sharp}(g)(z) \operatorname{av}_B w \,.$$

The proof is completed by combining these estimates. ∎

Lemma 7. Let $G(y,t) = g * \phi_t(y)$, and suppose $w \in A_\infty$. Then, given a ball B and $0 < s_2 < 1$, there are constants $a_2, \bar{s}_2$ independent of g and B such that for all $\lambda > 0$,

$$|\mathcal{E}_1| = |\{y \in B : \operatorname{av}_B w M_{1,\nu}^{\sharp}(g)(y) > a_2\lambda\}| < \bar{s}_2|B| \quad \text{implies}$$

$$|\mathcal{E}_2| = |\{y \in B : M_1^{0,2h}(G - (p_{3B}(g) * \phi_t), y) > \lambda\}| < s_2|b| \,.$$

Proof. Let

$$\mathcal{E}_3 = \{y \in B : \text{there is } B' \subseteq 3B, \, y \in B'$$

$$\text{and } \frac{1}{|B'|} \int_{B'} |p_{B'}(g)(y) - p_{3B}(g)(y)| \, dy > \alpha\lambda\} \,,$$

where $\alpha > 0$ is a constant to be chosen shortly, and $\mathcal{E}_4 = \{y \in B : \text{there is } B' \text{ so that } y \in B' \text{ and } \operatorname{av}_{B'} w > \delta \operatorname{av}_B w\}$, and where δ is also a constant to be chosen. First observe that $|\mathcal{E}_4|$ can be made arbitrarily small so long as δ is large enough. Indeed, to each y in $\mathcal{E}_4$ assign $B' \subseteq 3B$ with $\operatorname{av}_{B'} w > \delta \operatorname{av}_B w$. Now, by a familiar covering argument, there is a countable pairwise disjoint subfamily $\{B'_j\}$ such that $|\mathcal{E}_4| \leq c \sum_j |B'_j| \leq c\delta^{-1} \sum_j \nu(B'_j)(\operatorname{av}_B w)^{-1} \leq c\delta^{-1}|b|$, and the is true. Whence, if $\mathcal{E}_2 \subseteq \mathcal{E}_1 \cup \mathcal{E}_4$ we are done, and we can assume that this is not the case. Also, observe that if $y \in B \setminus (\mathcal{E}_1 \cup \mathcal{E}_3 \cup \mathcal{E}_4)$, then $M_1^{0,2h}(G - (p_{3B}(g) * \phi_t), y) < \lambda$. To see this, for each such y, let $(z,t) \in \Gamma_1^{0,2h}(y)$ and $B_1 = B(y,t)$. In this case, there are constants c_1, c_2 such that if $\psi(u) = c_1 e^{-c_2|u|^2}$, then $\phi_t(z-u) \leq \psi_t(y-u)$ for all $u \in R^n$. Consequently,

$$|G(z,t) - p_{3B}(g) * \phi_t(z)|$$

$$\leq (|g - p_{B_1}(g)|) * \psi_t(y) + (|p_{B_1}(g) - p_{3B}(g)|) * \psi_t(y)$$

$$= G_1(y,t) + G_2(y,t) \,,$$

say. An argument identical to that of the proof of Lemma 6 gives that $G_1(y,t) \leq c \operatorname{av}_B w M_{1,\nu}^{\sharp} g(y)$. Furthermore, since $y \notin \mathcal{E}_1 \cup \mathcal{E}_4$, this expression in turn is dominated by $c\delta \operatorname{av}_B w M_{1,\nu}^{\sharp} g(y) \leq c\delta a_2 \lambda < \lambda/2$, provided a_2 is small enough (the choice depending on δ of course.)

Now, recall that polynomials satisfy the following property: There exists a constant K such that $\sup_{2^j B_1} |p| \leq K^j \frac{1}{|B_1|} \int_{B_1} |p(u)| \, du$. Thus, if we pick $p(z) =$

$p_{B_1}(g)(z) - p_{3B}(g)(z)$, we have that $G_2(y,t)$ equals

$$\int_{B_1} |p(z)| \psi_t(y-z)\, dz + \sum_{j=0}^{\infty} \int_{2^{j+1}B_1 \setminus 2^j B_1} |p(z)| \psi_t(y-z)\, dz$$

$$\leq \frac{c}{t^n} \int_{B_1} |p(z)|\, dz + c \sum_{j=0}^{\infty} \left(\sup_{z \in 2^{j+1}B_1} |p(z)| \right) (2^{j+1}t)^n \frac{1}{t^n} \psi(2^{j+1})$$

$$\leq c\alpha\lambda + c \sum_{j=0}^{\infty} (K2^n)^{j+1} \left(\frac{1}{|B_1|} \int_{B_1} |p(z)|\, dz \right) \psi(2^{j+1})$$

$$\leq c\alpha\lambda + c\alpha\lambda \leq \lambda/2,$$

provided that α is small enough. Thus, $M_1^{0,2h}(F,y) < \lambda$ whenever y belongs to $B \setminus (\mathcal{E}_1 \cup \mathcal{E}_3 \cup \mathcal{E}_4)$. The assertion of the lemma follows once we estimate $|\mathcal{E}_3 \setminus \mathcal{E}_1|$, assuming there is a point $\overline{y} \in B \setminus (\mathcal{E}_1 \cup \mathcal{E}_4)$. To this end, to each $y \in \mathcal{E}_3 \setminus \mathcal{E}_1$ we assign a ball $B' \subseteq 3B$ such that

$$\frac{1}{|B'|} \int_{B'} |p'(z)|\, dz > \alpha\lambda, \quad p'(z) = p_{B'}(g)(z) - p_{3B}(g)(z).$$

By a familiar covering argument it follows that

$$|\mathcal{E}_3 \setminus \mathcal{E}_1| \leq c \sum_j |B_j'| \leq \frac{c}{\alpha\lambda} \sum_j \int_{B_j'} |p'(z)|\, dz$$

$$\leq \frac{c}{\alpha\lambda} \sum_j \int_{B_j'} |p_{B_j'}(g)(z) - g(z)|\, dz + \frac{c}{\alpha\lambda} \int_{B_j'} |g(z) - p_{3B}(g)(z)|\, dz$$

$$\leq \frac{c}{\alpha\lambda} \sum_j \nu(B_j') M_{1,\nu}^{\sharp} g(\overline{y}_j) + \frac{c}{\alpha\lambda} \int_{3B} |g(z) - p_{3B}(g)(z)|\, dz,$$

where $\overline{y}_j \notin \mathcal{E}_1$, and the B_j''s are pairwise disjoint. Thus, this last expression does not exceed

$$\frac{c}{\alpha\lambda} a_2 \lambda \sum_j |B_j'| + c\alpha\lambda \, \mathrm{av}_B w M_{1,\nu}^{\sharp} g(\overline{y})|B| \leq ca_2|B| + \frac{ca_2}{\alpha\lambda}|b| < ca_2|B|,$$

which can be made as small as we want provided a_2 is small enough. ∎

Corollary 8. Let g, G and w be as in Lemma 7. Given $0 < s_3 < 1$, there are constants $a_3, \overline{s}_3$ such that for all $\lambda > 0$,

$$|\{y \in B : M_{1,\nu}^{\sharp} g(y) > a_3\lambda\}| < \overline{s}_3|B| \quad \text{implies}$$

$$|\{y \in B : M_1^{0,2h}(G - (p_{3B} * \phi_t), y) > \lambda w(y)\}| < s_3|B|.$$

Proof.   We combine Lemma 7, with λ there replaced by $\mathrm{av}_B w\lambda$, and the observation in Lemma 5 to the effect that $|\{y \in B : w(y) < b\,\mathrm{av}_B w\}| < cb^\eta |B|$ for some $\eta > 0$ when $w \in A_\infty$, to obtain the desired conclusion at once. ■

Now, back to the proof of Theorem 3. Set $F(y,t) = f * \phi_t(y), F_1(y,t) = t|\nabla F(y,t)|, G(y,t) = g * \phi_t(y)$ and $G_1(y,t) = t|\nabla G(y,t)|$. To each ball $B = B(x,h)$ and $0 < s < 1$ we assign a number $A(B,s) = A_B$ as follows: Let A_f, A_g be such that

$$|\{y \in B : S_1^{0,h}(F_1,y) > A_f\}| < s|B|$$

and

$$|\{y \in B : S_1^{0,h}(G_1,y) > A_g w(y)\}| < s|B|\,,$$

put $A_B = \inf(A_f A_g)$, where A_f, A_g are defined above, and set

$$H(x) = \sup_{B \supseteq \{x\}} A_B\,.$$

By Lemmas 4 and 5 and Corollary 8, it readily follows that there are constants c and $0 < s_1, s_2 <$ such that

$$H(x) \leq M_{0,s_1}\left(M_1(F)M_{0,s_2}(M_{1,\nu}^\sharp g)\right)(x)\,.$$

We still need one more function, namely the stopping time $h(x)$ given by

$$h(x) = \sup\{h > 0 : S_1^{0,h}(F_1,x)S_1^{0,h}(G_1,x) \leq w(x)H(x)\}\,.$$

Now, if B is a ball of radius h and $x \in B$, pick A_f, A_g such that $A_B \sim A_f A_g$, and let $\mathcal{A}_f = \{y \in B : S_1^{0,h}(F_1,y) > A_f\}$ and $\mathcal{A}_g = \{y \in B : S_1^{0,h}(G_1,y) > w(y)A_g\}$; then

$$|\mathcal{A}_f| < s|B| \quad \text{and} \quad |\mathcal{A}_g| < s|B|\,.$$

Moreover, since $\{y \in B : S_1^{0,h}(F_1,y)S_1^{0,h}(G,y) > w(y)A_f A_g\} \subseteq \mathcal{A}_f \cup \mathcal{A}_g$, it follows that

$$|\{y \in B : S_1^{0,h}(F_1,y)S_1^{0,h}(G_1,y) > w(y)A_B\}| < 2s|B|\,.$$

Since by the definition we get that $\inf_{y \in B} H(y) \geq A_B$, we have

$$|\{y \in B : S_1^{0,h}(F_1,y)S_1^{0,h}(G_1,y) > w(y)H(y)\}| < 2s|B|\,.$$

This inequality implies that

$$|\{y \in B : h(y) < h\}| < 2s|B|\,.$$

132

Indeed, if y is such that $h(y) < h$, note that $S_1^{0,h}(F_1,y)S_1^{0,h}(G_1,y) > w(y)H(y)$. Thus, $|\{y \in B : h(y) \geq h\}| \geq |B|/2$ whenever $0 < s < 1/4$. By Fubini's theorem it follows that

$$
\left| \int_{R^n} f(y)g(y)\,dy \right| = c \left| \int_{R^n} \int_0^\infty \langle t\nabla F(y,t) \cdot t\nabla G(y,t) \rangle \frac{dt}{t}\,dy \right|
$$

$$
\leq c \int_{R^n} \int_0^\infty \int_{\{y:\,|x-y|<t<h(x)\}} t^{-n} F_1(y,t)G_1(y,t)dy \frac{dt}{t}\,dx
$$

$$
\leq c \int_{R^n} S_1^{0,h(x)}(F_1,x)S_1^{0,h(x)}(G_1,x)\,dx \leq c \int_{R^n} H(x)w(x)\,dx
$$

$$
\leq c \int_{R^n} M_{0,s_1}\left(M_1(F)M_{0,s_2}(M_{1,\nu}^\sharp g) \right)(x)w(x)\,dx .
$$

By Lemma 1 in Chapter III, with $\Phi(t) = t$ there, it follows that the above expression is dominated by

$$
\int_{R^n} M_1(F,x)M_{0,s_2}(M_{1,\nu}^\sharp g)(x)w(x)\,dx .
$$

So, to complete the proof it suffices to show that

$$
M_{0,s_2}(M_{1,\nu}^\sharp g)(x) \leq c M_{1,\nu}^\sharp g(x) .
$$

First observe that, by the results in Chapter III, if $w \in A_\infty$, then $M_{0,s}g(x) \leq c M_{0,\nu,s_1}g(x)$ where s_1 depends on s and on the A_p class of w. Whence, it is enough to show that $M_{0,\nu,s}(M_{1,\nu}^\sharp g)(x) \leq c M_{1,\nu}^\sharp g(x)$; this is not hard. Fix x and let $B \supset \{x\}$. We begin by observing that there is a constant α, independent of B, such that if $\mathcal{E} = \{y \in B : M_{1,\nu}^\sharp g(y) > \alpha M_{1,\nu}^\sharp g(x)\}$, then $\nu(E) < s\nu(B)$. Indeed, let $y \in \mathcal{E}$; α will be chosen shortly. Then there is a ball $B_y \supset \{y\}$ such that

$$
\inf_c \frac{1}{\nu(B_y)} \int_{B_y} |g(z) - c|\,dz > \alpha M_{1,\nu}^\sharp g(x) .
$$

We claim that $B_y \subseteq 2B$. For, if not, any point in B, and in particular x, belongs to $4B_y$. It then follows that

$$
\alpha M_{1,\nu}^\sharp g(x) \leq \frac{1}{\nu(B_y)} \int_{B_y} |g(z) - p_{B_y}(g)(z)|\,dz
$$

$$
\leq \frac{\nu(4B_y)}{\nu(B_y)} M_{1,\nu}^\sharp g(y) \leq c M_{1,\nu}^\sharp g(x) ,
$$

which is a contradiction if α is large enough; this is one condition α must meet. Let now $\{B_{y_i}\} = \{B_i\}$ be a pairwise disjoint countable subfamily of the B_y's such that $\nu(\mathcal{E}) \leq c \sum_i \nu(B_i)$. We then have

$$
\begin{aligned}
\sum_i \nu(B_i) &\leq \frac{1}{\alpha M_{1,\nu}^{\sharp} g(x)} \sum_i \int_{B_i} |g(z) - p_{B_i}(g)(z)| \, dz \\
&\leq \frac{1}{M_{1,\nu}^{\sharp} g(x)} \sum_i \int_{B_i} |g(z) - p_{2B}(g)(z)| \, dz \\
&\leq \frac{1}{M_{1,\nu}^{\sharp} g(x)} \int_{2B} |g(z) - p_{2B}(g)(z)| \, dz \\
&\leq \frac{1}{M_{1,\nu}^{\sharp} g(x)} \nu(2B) M_{1,\nu}^{\sharp} g(x) < \frac{c}{\alpha} \nu(B) \,.
\end{aligned}
$$

We choose now α so large that $c/\alpha < s$; α is independent of x and B. Whence,

$$
M_{0,\nu,s}^{\sharp}(M_{1,\nu}^{\sharp} g)(x) \leq \alpha M_{1,\nu}^{\sharp} g(x) \,,
$$

and the proof is complete. ∎

Sources and Remarks. The proof of Theorem 3 follows along the lines of its unweighted version, due to J.-O. Strömberg [1979a]. On the other hand, the proof using the atomic decomposition, as in Proposition 1, follows that of R. Fefferman for the dyadic case and that of E. Stein for the general case.

CHAPTER

X

Duality

In this chapter we describe the space $(H_w^p(R^n))^*$ of bounded linear functionals on $H_w^p(R^n)$. Since the case $0 < p < 1$ is better known, we will be mainly concerned with the case $p \geq 1$; then, of course, $H_w^p(R^n)$ is a Banach space.

First recall that, as established in the proof of Lemma 2 in Chapter VIII, for any $p > 0$, the $H_w^p(R^n)$-quasi norm induces a topology on $H_w^p(R^n)$ which is stronger than the weak-topology it inherits from the space $S'(R^n)$ of tempered distributions. This is a consequence of the estimate $|\langle f, \phi \rangle| \leq c_\phi \|f\|_{H_w^p}$, ϕ in $S(R^n)$, established there. Thus, every $\phi \in S(R^n)$ defines a continuous linear functional ℓ_ϕ, say, on $H_w^p(R^n)$ by means of

$$\ell_\phi(f) = \langle f, \phi \rangle, \quad f \in H_w^p(R^n).$$

On the other hand, it is not hard to see that every continuous linear functional ℓ on $H_w^p(R^n)$ can be represented on a suitable dense class by a tempered distribution $g_\ell \in S'(R^n)$, i.e.,

$$\ell(f) = \langle g_l, f \rangle, \quad f \in H_w^p(R^n) \cap S(R^n).$$

To see this, observe that there exists an integer $N > 0$ and a finite number of Schwartz norms $\| \cdot \|_k$ depending only on p and w such that if $f \in S(R^n)$, then f belongs to $H_w^p(R^n)$ if and only if $\int_{R^n} x^\alpha f(x) \, dx = 0$ for $|\alpha| \leq N$, and $\|f\|_{H_w^p} \leq c \sum_k \|f\|_k$; we do not prove this statement at this time for it is a particular case of results proved later in the chapter.

So, having framed $S(R^n) \subseteq (H_w^p(R^n))^* \subseteq S'(R^n)$, we seek a closer identification of the space of functionals. In particular, we want to replace the Schwartz class in the arguments above by a space, to be denoted by $\mathcal{A}$, which is as large as possible, and then to identify $\ell \in (H_w^p(R^n))^*$ on $H_w^p(R^n) \cap \mathcal{A}$ with an element $g_\ell \in \mathcal{A}^*$, the dual space to $\mathcal{A}$. Moreover, elements in $\mathcal{A}^*$ should be easily described as either functions or measures.

If $\mathcal{A}$ contains the finite linear combinations of atoms, or if at least the representation of $\ell \in (H_w^p(R^n))^*$ by $g \in \mathcal{A}^*$ is valid for finite sums of atoms, we hope to derive estimates for the sharp maximal function $M_{1,\nu}^{\sharp,\mathcal{P}} g_\ell$. In fact, the estimates we obtain are

1. If $\|\ell\|$ denotes the norm of ℓ as a functional and $0 < p < 1$, for an appropriate A we have

$$\|M_{1,\nu A}^{\sharp,\mathcal{P}} g_\ell\|_{L^\infty} \le c\|\ell\|.$$

2. In the notation of 1. above, if $1 \le p < \infty$ and $1/p + 1/p' = 1$, and if $\mathcal{P}$ denotes an appropiate class of polynomials, the degree depends on p, ν, we have

$$\|M_{1,\nu}^{\sharp,\mathcal{P}} g_\ell\|_{L_w^{p'}} \le c\|\ell\|.$$

The appropriate A in condition 1 above is determined from the basic inequality. Also, if the left-hand side in 1 above is finite for a locally integrable function g, it follows from the basic inequality that such g's define bounded linear functionals on $H_w^p(R^n)$. Similarly, when $1 \le p < \infty$ and $f \in \hat{\mathcal{D}}_0$, by the basic inequality and Hölder's inequality we have

$$\left| \int_{R^n} f(x)g(x)\,dx \right| \le c \int_{R^n} M_1(F_\phi, x) M_{1,\nu}^{\sharp,\mathcal{P}} g(x) w(x)\,dx$$
$$\le c\|f\|_{H_w^p} \|M_{1,\nu}^{\sharp,\mathcal{P}} g\|_{L_w^{p'}}.$$

Thus, any locally integrable function g such that $\|M_{1,\nu}^{\sharp,\mathcal{P}} g\|_{L_w^{p'}} < \infty$ induces a linear functional on $\hat{\mathcal{D}}_0$ that is bounded in the $H_w^p(R^n)$-norm on the dense class $\hat{\mathcal{D}}_0$ and which can, therefore, be extended continuously to a functional $\ell \in (H_w^p(R^n))^*$. If also $\hat{\mathcal{D}}_0 \subseteq \mathcal{A}$, then there is a $g_\ell \in \mathcal{A}^*$ such that

$$\int_{R^n} f(x)g(x)\,dx = \int_{R^n} f(x)g_\ell(x)\,dx, \quad \text{all } f \in \hat{\mathcal{D}}_0,$$

and $g(x) - g_\ell(x)$ is a polynomial.

Our goal is to obtain a description of the space of functionals in terms of $\mathcal{A}$ along the following lines: If $\mathcal{P}_k$ denotes the class of polynomials of degree less that or equal to k, $k = 0, 1, \ldots$, then,

(i) We have, with $N = N(p, w)$,

$$H_w^p(R^n) \cap \mathcal{A} = \left\{ f \in \mathcal{A} : \int_{R^n} f(x)p(x)\,dx = 0, p \in \mathcal{P}_N \right\}.$$

(ii) $\mathcal{P}_N \subset \mathcal{A}^*$.

(iii) The space $(H_w^p(R^n))^*$ is in a one-to-one correspondence with $\{g \in \mathcal{A}^* : M_{1,\nu}^{\sharp,\mathcal{P}_m} g \in L_w^{p'}(R^n)\}/\mathcal{P}_N$, where $N \leq m = m(p,w)$. The correspondence $g \to \ell \in (H_w^p(R^n))^*$ is given, for $f \in H_w^p(R^n) \cap \mathcal{A}$, by

$$\ell(f) = \int_{R^n} f(x)g(x)\,dx, \quad \text{and} \quad \|\ell\|_* \sim \|M_{1,\nu}^{\sharp,\mathcal{P}_m} g\|_{L_w^{p'}}.$$

Observe that under these circumstances, $x^\alpha \notin \mathcal{A}^*$ when $|\alpha| > N$. Indeed, if $x^\alpha \in \mathcal{A}^*$, on the one hand it induces a linear functional which vanishes on the dense subspace $\hat{\mathcal{D}}_0$ and, on the other hand, this linear functional is not identically zero since, by (i) above, there are functions ϕ in $H_w^p(R^n) \cap \mathcal{A}$ with $\int_{R^n} \phi(x)x^\alpha\,dx \neq 0$; we are assuming $\mathcal{S} \subseteq \mathcal{A}$.

Summing up, we search for a space $\mathcal{A}$ with the following properties: There is an integer $N = N(p,w)$ so that

1. $f \in H_w^p(R^n) \cap \mathcal{A}$ if and only if $f \in \mathcal{A}$ and

$$\int_{R^n} x^\alpha f(x)\,dx = 0, \quad |\alpha| \leq N.$$

2. There is a constant c independent of f such that

$$\|f\|_{H_w^p} \leq c\|f\|_{\mathcal{A}}, \quad \text{all } f \in H_w^p(R^n) \cap \mathcal{A}.$$

3. $\mathcal{A}^*$ contains all polynomials of degree less than or equal to N, but no polynomial of degree greater than N.

4. $\mathcal{A}$ contains $\hat{\mathcal{D}}_0$, and all finite linear combinations of atoms.

5. $\mathcal{A}^*$ can be described in an explicit way.

The atoms we use are (q, N_1) atoms, with the choice $q = \infty$ when we only assume that the measure ν is doubling. If in addition $w \in A_\infty$, and in this case it satisfies a reverse Hölder's condition RH_r for some $r > 1$, then we may pick $q = pr'$. When $q = \infty$, it is not always possible to satisfy simultaneously conditions 4 and 5 above, because the dual of a space given by an L^∞-norm is not easily described. This difficulty is overcome by requiring that $\mathcal{A}$ contain all finite linear combinations of continuous atoms.

In addition to the choice of q described above, we associate with the weight w the set

$$I_w = \left\{ t > 0 : \int_{R^n} (1 + |x|)^{-tp} w(x)\,dx < \infty \right\};$$

I_w is an interval of the form (t_0, ∞) or $[t_0, \infty)$ for some $t_0 > 0$. $\mathcal{A}$ will then be described in terms of q and t_0.

Let N' be the largest integer such that $n + N \notin I_w$, i.e., $n + N + 1$ is the smallest integer that belongs to I_w. We consider two cases, to wit,

Case 1. $t_0 \neq n + N + 1$.

Case 2. $t_0 = n + N + 1$.

We begin by discussing Case 1. Pick t_1 so that $t_0 < t_1 < n + N + 1$. If $1 < q < \infty$, we let $\mathcal{A}$ consist of those locally integrable functions f which belong to $L_u^q(R^n)$, where $u(x) = (1 + |x|)^{t_1 q - n}$, and $\mathcal{A}^*$ of those locally integrable g's that belong to $L_v^{q'}(R^n)$, where $v(x) = (1 + |x|)^{-t_1 q' + (n/(q-1))}$. Specifically, let

$$\mathcal{A} = \left\{ f \in L_{\text{loc}}^1(R^n) : \left(\int_{R^n} |f(x)|^q (1 + |x|)^{t_1 q - n} \, dx \right)^{1/q} = \|f\|_{\mathcal{A}}^q < \infty \right\},$$

and

$$\mathcal{A}^* = \left\{ g \in L_{\text{loc}}^1(R^n) : \left(\int_{R^n} |g(x)|^{q'} (1 + |x|)^{-t_1 q' + (n/(q-1))} \, dx \right)^{1/q'} < \infty \right\}.$$

On the other hand, if $q = \infty$, let

$$\mathcal{A} = \{ f \in C(R^n) : |f(x)| = o(|x|^{-t_1}) \text{ as } |x| \to \infty \}$$

and set $\|f\|_{\mathcal{A}} = \sup_{x \in R^n} |f(x)|(1 + |x|)^{t_1}$, and

$$\mathcal{A}^* = \left\{ \text{locally finite measures } \mu : \int_{R^n} (1 + |x|)^{-t_1} |d\mu|(x) = \|\mu\|_{\mathcal{A}^*} < \infty \right\}.$$

As for the Case 2, note that

$$\int_{R^n} (1 + |x|)^{-t_0 p} w(x) \, dx < \infty.$$

We can then find a very slowly increasimg function $\varrho : [0, \infty) \to [1, \infty)$, that satisfies the following conditions:

$$\int_{R^n} (1 + |x|)^{-t_0 p} \varrho(|x|)^p w(x) \, dx < \infty$$

and

$$\varrho(s) = 1 \text{ for } 0 \leq s \leq 1, \quad \text{and} \quad \varrho(s) \to \infty \text{ as } s \to \infty.$$

In addition we require that

$$0 \leq \varrho(ts) - \varrho(s) \leq \ln t, \quad t \geq 1.$$

Lemma 1. Suppose ϱ is as above. We then have

(a) $\sup_{s>0} (1 + (s/t))^{-1} \varrho(s) \leq c \varrho(t), t > 0$.

(b)

$$\sup_{|s-t|<(1+t)/2} \varrho(s) \leq c \inf_{|s-t|<(1+t)/2} \varrho(s), \quad t > 0.$$

Proof. We show (a) first. When $s \leq t$ we have $(1 + (s/t))^{-1}\varrho(s) \leq \varrho(s) \leq \varrho(t)$, and when $s \geq t$ we have

$$(1 + (s/t))^{-1}\varrho(s) \leq (1 + (s/t))^{-1}(\varrho(t) + \ln(s/t))$$
$$\leq \varrho(t) + (1 + (s/t))^{-1}\ln(s/t) \leq \varrho(t) + c \leq c\varrho(t).$$

As for (b), if $0 \leq t \leq 3$, then $0 \leq s \leq 5$. Thus, both sides of the inequality assume values in the interval $[\varrho(0), \varrho(5)] \subseteq [1, \ln 5]$. On the other hand, if $t > 3$, then $t/3 < s < 5t/3$. Whence, if $\varrho(t) > 4$ we have

$$\varrho(t) - \ln 3 \leq \varrho(s) \leq \varrho(t) + \ln(5/3).$$

Therefore, $\varrho(t)/2 \leq \varrho(s) \leq 2\varrho(t)$. Also, if $\varrho(t) \leq 4$, we have $1 \leq \varrho(s) \leq \varrho(t) + \ln(5/3)$. Whence, $\varrho(t)/4 \leq \varrho(s) \leq 2\varrho(t)$, and (b) follows. ∎

We are now in a position to define $\mathcal{A}$ when $t_0 = n + N + 1 \in I_w$. Suppose first that $q < \infty$. Let

$$d\mu_1(x) = (1 + |x|)^{t_0 - n}\varrho(|x|)^{-1}dx, \quad d\mu_q(x) = (1 + |x|)^{t_0 q - n}dx,$$

and set

$$\mathcal{A} = L^1(\mu_1) \cap L^q(\mu_q), \quad \text{with norm} \quad \|f\|_{\mathcal{A}} = \|f\|_{L^1(\mu_1)} + \|f\|_{L^q(\mu_q)}.$$

Now, if $d\mu_{q'}(x) = (1 + |x|)^{-(t_0 - (n/q))q'}dx$, we can describe $\mathcal{A}^*$ as follows: Let

$$\|g\|_{\mathcal{A}_1^*} = \sup_x (1 + |x|)^{n - t_0}\varrho(|x|)|g(x)|,$$

and set

$$\mathcal{A}^* = L^{q'}(\mu_{q'}) \cup \{g \in L^1_{\text{loc}} : \|g\|_{\mathcal{A}_1^*} < \infty\},$$

with norm

$$\|g\|_{\mathcal{A}^*} = \inf_{g = g_1 + g_2} \left\{\|g_1\|_{\mathcal{A}_1^*} + \|g_2\|_{L^{q'}(\mu_{q'})}\right\}.$$

Finally, when $q = \infty$ we put

$$\mathcal{A} = \{f \in C(R^n) : |f(x)| = o(|x|^{-t_0}) \text{ as } |x| \to \infty\} \cap L^1(\mu_1),$$

and

$$\mathcal{A}^* = \{g : \|g\|_{\mathcal{A}_1^*} < \infty\}$$

$$\bigcup \left\{\text{Borel measures } \nu_2 : \int_{R^n} (1 + |x|)^{-t_0}|d\nu_2|(x) = \|\nu_2\|_{\mathcal{A}_2^*} < \infty\right\}.$$

In this case we set

$$\|f\|_{\mathcal{A}} = \|f\|_{L^1(\mu_1)} + \sup_x (1 + |x|)^{t_0}|f(x)|,$$

and

$$\|\nu\|_{\mathcal{A}^*} = \inf_{d\nu(x) = g_1(x)dx + d\nu_2(x)} (\|g_1\|_{\mathcal{A}_1^*} + \|\nu_2\|_{\mathcal{A}_2^*}).$$

With these definitions out of the way, we begin with the following observation.

Lemma 2. $(1 + |x|)^N$ and x^α with $|\alpha| \le N$ belong to $\mathcal{A}^*$; $(1 + |x|)^{N+1}$ and polynomials of degree greater than or equal to $N + 1$ do not.

Proof. It suffices to verify the claims for $(1 + |x|)^N$ and $(1 + |x|)^{N+1}$. In case 1, this amounts to a verification that is left to the reader. In case 2, note that

$$(1 + |x|)^N (1 + |x|)^{-t_0} = (1 + |x|)^{-n-1} \in L^1(R^n) \cap L^q(R^n),$$

and so it defines a bounded measure. On the other hand,

$$(1 + |x|)^{n-t_0} \varrho(|x|)(1 + |x|)^{N+1} = \varrho(|x|) \notin L^\infty(R^n),$$

and

$$(1 + |x|)^{-t_0}(1 + |x|)^{N+1} dx = (1 + |x|)^{-n} dx$$

is not a finite measure. Also,

$$(1 + |x|)^{-(t_1 - (n/q))}(1 + |x|)^{N+1} = (1 + |x|)^{-n/q'} \notin L^{q'}(R^n).$$

This completes the proof. ∎

As for $H_w^p(R^n) \cap \mathcal{A}$, we have

Lemma 3. Let $f \in H_w^p(R^n) \cap \mathcal{A}$. Then, $\int_{R^n} x^\alpha f(x)\, dx = 0$ when $|\alpha| \le N$.

Proof. Let $|\alpha| \le N$, and assume that the Schwartz function ϕ coincides with x^α in a neighborhood of the origin. Then,

$$|\phi_t(x) - t^{-n-|\alpha|} x^\alpha| \le \begin{cases} 0 & \text{if } |x| \le ct \\ ct^{-n-|\alpha|} |x|^{|\alpha|} & \text{if } |x| > ct. \end{cases}$$

Since, by Lemma 2, $|x|^{|\alpha|} \in \mathcal{A}^*$, it follows that

$$|f * \phi_t(0)| = \left| \int_{R^n} f(x)\phi_t(-x)\, dx \right|$$

$$\ge \left| t^{-n-|\alpha|} \int_{R^n} x^\alpha f(x)\, dx \right| - \left| \int_{R^n} (\phi_t(-x) - t^{-n-|\alpha|}(-x^\alpha)) f(x)\, dx \right|$$

$$\ge t^{-n-|\alpha|} \left(\left| \int_{R^n} x^\alpha f(x)\, dx \right| - \int_{\{|x|>ct\}} |x|^{|\alpha|} |f(x)|\, dx \right).$$

The last integral above goes to 0 as $t \to \infty$. Thus, if $\int_{R^n} x^\alpha f(x)\, dx = c_\alpha \ne 0$, it follows that

$$|f * \phi_t(0)| \ge |c_\alpha| t^{-n-|\alpha|}/2, \quad t \text{ large.}$$

Whence, if $F(x,t) = f * \phi_t(x)$,

$$M_1(F,x) \geq c|x|^{-(n+|\alpha|)}, \quad |x| \text{ large}$$

and since $f \in H_w^p(R^n)$,

$$I = \int_{\{|x|>c\}} |x|^{-(n+|\alpha|)p} w(x)\, dx \leq c\|M_1(F)\|_{H_w^p} < \infty.$$

However, this is a contradiction, since $(n+|\alpha|) \notin I_w$ for $|\alpha| \leq N$ and consequently $I = \infty$. This completes the proof. ∎

Next we show that, under weak additional assumptions, elements in $\mathcal{A}$ are also in $H_w^p(R^n)$.

Lemma 4. Suppose $f \in \mathcal{A}$ and $\int_{R^n} x^\alpha f(x)\, dx = 0$ for $|\alpha| \leq N$. Then f belongs to $H_w^p(R^n)$ and $\|f\|_{H_w^p} \leq c\|f\|_{\mathcal{A}}$.

Proof. Let ϕ be a Schwartz function supported in $B(0,1)$ with nonvanishing integral and set $F(x,t) = f * \phi_t(x)$. We write $M_1(F,x) = M_1'(F,x) + M_1''(F,x)$, where

$$M_1'(F,x) = \sup_{\substack{|x-y|<t \\ (1+|x|)/5<t}} |F(y,t)|, \quad \text{and} \quad M_1''(F,x) = \sup_{\substack{|x-y|<t \\ t\leq(1+|x|)/5}} |F(y,t)|.$$

We consider $M_1'(F,x)$ first. Let $P_{t,y}$ denote the Nth order Taylor polynomial of the function $z \to \phi_t(y-z)$ at $z=0$. We then have

$$|\phi_t(y-z) - P_{t,y}(z)| \leq ct^{-n-N-1}|z|^{N+1}/(1+(|z|/t)),$$

uniformly on $|y| \leq 6t$.

Note that $t > (1+|x|)/5$ and $t > |x-y|$ imply that $|y| < 6t$. Now, since $\int_{R^n} f(z)P_{t,y}(z)\, dz = 0$, for $|y| < 6t$ it follows that

$$|F(y,t)| = \left| \int_{R^n} f(z)(\phi_t(y-z) - P_{t,y}(z)\, dz \right|$$
$$\leq ct^{-(n+N+1)} \int_{R^n} |f(z)||z|^{N+1}/(1+(|z|/t))\, dz. \qquad (1)$$

In case 1, let t_2 satisfy $n + N \leq t_0 < t_2 < t_1$. For $t \geq c$ we have

$$t^{-(n+N+1)}|z|^{N+1}/(1+(|z|/t)) \leq c(1+|z|)^{t_2-n}t^{-t_2},$$

and consequently, from (1) it follows that

$$|F(y,t)| \leq ct^{-t_2} \int_{R^n} |f(z)|(1+|z|)^{t_2-n} dz$$

$$\leq ct^{-t_2} \|f\|_A \left(\int_{R^n} (1+|z|)^{(t_2-t_1)q'-n} dz \right)^{1/q'} = ct^{-t_2} \|f\|_A.$$

If now $t > (1+|x|)/5$, we conclude that

$$M_1'(F,x) \leq c(1+|x|)^{-t_2} \|f\|_A.$$

Moreover, since $t_2 \in I_w$, $(1+|x|)^{-t_2}$ is in $L_w^p(R^n)$ and consequently, $\|M_1'(F)\|_{L_w^p} \leq c\|f\|_A$.

In case 2 we use Lemma 1 instead to get

$$t^{-(n+N+1)}|z|^{N+1}/(1+(|z|/t)) \leq c\varrho(t)t^{-(n+N+1)}|z|^{N+1}\varrho(|z|)^{-1}.$$

Thus, from (1) we get

$$|F(y,t)| \leq c\varrho(t)t^{-(n+N+1)} \int_{R^n} |f(z)|(1+|z|)^{t_0-n}\varrho(|z|)^{-1} dz$$

$$\leq c\varrho(t)t^{-t_0}\|f\|_A,$$

and since $t > c(1+|x|)$ we conclude that $M_1'(F,x) \leq c\varrho(|x|)(1+|x|)^{-t_0}\|f\|_A$. Also, since $t_0 \in I_w$, we have $(1+|x|)^{-t_0}\varrho(|x|) \in L_w^p(R^n)$, and consequently, $\|M_1'(F)\|_{L_w^p} \leq c\|f\|_A$.

We pass now to estimate $M_1''(F,x)$. Since ϕ is supported in $\{z : |z| < 1\}$, $\phi_t(y-\cdot)$ is supported in $\{z : |z-y| < t\}$, and if also $t < (1+|x|)/5$ and $|y-x| < t$, then the support is contained in the set $\{z : |z-x| < (1+|x|)/2\}$. Now, for all real numbers s,

$$\frac{1}{c} \sup_{|z-y|<(1+|x|)/2} (1+|z|)^s \leq (1+|x|)^s \leq c \inf_{|z-y|<(1+|x|)/2} (1+|z|)^s. \qquad (2)$$

So, for any real number s and for $|y-x| < t < (1+|x|)/5$,

$$|F(y,t)| \leq (1+|x|)^{-s} \int_{R^n} |\phi_t(y-z)|(1+|z|)^s |f(z)| dz$$

$$\leq c(1+|x|)^{-s} M((1+|\cdot|)^s f)(x),$$

where M denotes the Hardy-Littlewood maximal function with respect to the Lebesgue measure on R^n. It then follows that

$$M_1''(F,x) \leq c(1+|x|)^{-s} M((1+|\cdot|)^s f)(x)$$

and consequently, by Hölder's inequality with indices r and r' we get

$$\|M_1''(F)\|_{L_w^p} \leq c\|w(1+|\cdot|)^{-s}\|_{L^r}^{1/p}\|M((1+|\cdot|)^s f)\|_{L^q}.$$

On account of the RH_r assumption on w and the L^q continuity properties of the Hardy-Littlewood maximal function, in case 1, for an appropriate choice of s, this last estimate yields that $\|M_1''(F)\|_{L_w^p} \leq c\|f\|_{\mathcal{A}}$.

To obtain the same estimate in case 2, it suffices to observe that the inequalities in (2) hold with $(1+|x|)^s$ replaced by $(1+|x|)^s \varrho(|x|)$. Thus, collecting the estimates for $M_1'(F)$ and $M_1''(F)$ we conclude that, as asserted, $\|f\|_{H_w^p} \leq c\|M(F)\|_{L_w^p} \leq c\|f\|_{\mathcal{A}}$. ∎

Along the lines of the argument outlined at the begining of the chapter we can now represent a linear functional $\ell \in (H_w^p(R^n))^*$, $1 \leq p < \infty$, on $H_w^p(R^n) \cap \mathcal{A}$ by means of an element g in $\mathcal{A}^*$, and this representation is unique modulo polynomials of degree less than or equal to N. Now, when $q = \infty$ and w is not an A_∞ weight, it is not known a-priori whether the measure g is locally absolutely continuous; we show next that this is the case. It is therefore possible to extend the representation of ℓ to $H_w^p(R^n) \cap \tilde{\mathcal{A}}$, where

$$\tilde{\mathcal{A}} = \{g : (1+|x|)^{t_0} g \in L^\infty(R^n)\}, \quad t_0 \neq n+N+1,$$

or

$$\tilde{\mathcal{A}} = \{g : (1+|x|)^{t_0} g \in L^\infty(R^n)\} \cap \{g : (1+|x|)^{t_0-n} \varrho(|x|)^{-1} g \in L^1(R^n)\},$$

when $t_0 = n+N+1$.

We begin, then, by showing the statement concerning absolute continuity.

Lemma 5. Let $\ell \in (H_w^p(R^n))^*$, $1 \leq p < \infty$, and suppose $\mu \in \mathcal{A}^*$ is such that

$$\ell(f) = \int_{R^n} f(x)\, d\mu(x), \quad \text{all } f \in H_w^p(R^n) \cap \mathcal{A}.$$

Then $d\mu(x) = g(x)dx$, where g is locally integrable.

Proof. Let Q be a fixed cube. We want to show that given $\varepsilon > 0$, there is $\delta > 0$ such that if $E \subset Q$ has Lebesgue measure less than or equal to δ, then $\mu(E) \leq \varepsilon$; we may assume that Q is the unit cube centered at the origin. Let $|\mu| = \mu_+ + \mu_-$, where μ_+ and μ_- denote the positive and negative parts of μ respectively, and let $c_1 = |\mu|(Q)$. If $\delta > 0$ is given, and if E is a measurable subset of Q, there is a continuous function g_δ such that $0 \leq g_\delta \leq 1$ and the set $G_\delta = \{x \in Q : g_\delta(x) \neq \chi_E(x)\}$ satisfies $|G_\delta| < \delta$ and $|\mu|(G_\delta) < \delta$. Thus, if also $|E| \leq \delta$, then $|\operatorname{supp} g| \leq 2\delta$.

Let ϕ be a continuous function, $0 \leq \phi \leq 1$, that vanishes off Q and so that ϕ equals 1 for $|x| \leq 1/2$. Then we can find a polynomial p_δ such that

$$\int_{R^n} (g_\delta(x) - p_\delta(x)\phi(x))x^\alpha \, dx = 0 \quad \text{for} \quad |\alpha| \leq N\,, \quad \text{and} \quad \|\phi p_\delta\|_{L^\infty} \leq c\delta\,.$$

Then $a = g_\delta - \phi p_\delta$ is an (∞, N) atom and $\|a\|_{H^p_w} \leq c$. In fact, as we now show in Lemma 6, $\|a\|_{H^p_w} = O(1)$ as $\delta \to 0$, uniformly for $E \subseteq Q$, $|E| < \delta$.

Lemma 6. Let ν be a doubling weighted measure with respect to the Lebesgue measure on R^n, $d\nu(x) = w(x)dx$, Q a cube of R^n, and $1 \leq p < \infty$, $1 < q < \infty$, $p \leq q$. Then there is a function $\sigma(t)$, defined for $t > 0$, such that $\lim_{t \to 0} \sigma(t) = 0$ and if $a \in H^p_w(R^n)$ is an atom supported in Q satisfying $\|a\|_{L^\infty} \leq 1$ and $\|a\|_{L^q} \leq \delta$, then $\|a\|_{H^p_w} \leq \sigma(\delta)$.

Proof. Since $a \in H^p_w(R^n)$, by Lemma 3, $\int_{R^n} x^\alpha a(x) \, dx = 0$ for $|\alpha| \leq N$. Therefore, since $\|a\|_{L^q} < \delta$, it readily follows that given a Schwartz function ψ,

$$M_1(a * \psi_t, x) \leq c\delta|x|^{-n-(N-1)}\,, \quad |x| > 2\,.$$

Thus,

$$\int_{\{|x|>2\}} |M_1(a * \psi_t, x)|^p w(x) \, dx \leq c\delta^p\,.$$

For $|x| \leq 2$ we use the estimates

$$\|M_1(a * \psi_t)\|_{L^\infty} \leq c\|a\|_{L^\infty} \leq c\,, \quad \text{and} \quad \|M_1(a * \psi_t)\|_{L^q} \leq c\|a\|_{L^q} \leq c\delta\,.$$

Let

$$\lambda(\delta) = \int_{\{|x| \leq 2\}} \chi_{\{w > \delta^{-1/2}\}}(x) w(x) \, dx\,;$$

then, $\lim_{\delta \to 0} \lambda(\delta) = 0$. Whence,

$$\int_{\{|x| \leq 2\}} M_1(a * \psi_t, x)^p w(x) \, dx$$

$$\leq \lambda(\delta)\|M_1(a * \psi_t)\|_{L^\infty}^p + \delta^{-p/2} \int_{\{|x| \leq 2\}} M_1(a * \psi_t, x)^p w(x) \, dx$$

$$\leq \lambda(\delta)\|M_1(a * \psi_t)\|_{L^\infty}^p + c\delta^{-p/2}\|M_1(a * \psi_t)\|_{L^q}^p$$

$$\leq c\lambda(\delta) + c\delta^{p/2}\,.$$

Set $\sigma(t) = (\lambda(t) + ct^{p/2} + t^p)^{1/p}$; clearly $\lim_{t \to 0} \sigma(t) = 0$. Also, from the above estimates it follows that

$$\|a\|_{H^p_w} \leq c\|M_1(a * \psi_t)\|_{L^p_w} \leq c\sigma(\delta)\,. \quad \blacksquare$$

Proof of Lemma 5, continued. For $p \leq q < \infty$, the atom $a = g_\delta - \phi p_\delta$ satisfies

$$\|a\|_{L^q} \leq \|g_\delta\|_{L^q} + \|\phi p_\delta\|_{L^q} \leq c(\delta^{1/q} + \delta),$$

and, from Lemma 6, we conclude that $\|a\|_{H^p_w} \leq c\sigma(c\delta^{1/q})$. Further, since a is continuous, $a \in H^p_w(R^n) \cap \mathcal{A}$, and

$$\left| \int_{R^n} a(x)\, d\mu(x) \right| = |\ell(a)| \leq \|\ell\| \, \|a\|_{H^p_w} \leq c\|\ell\|\sigma(c\delta^{1/q}).$$

On the other hand,

$$\int_{R^n} a(x)\, d\mu(x) = \int_{R^n} \chi_E(x)\, d\mu(x)$$
$$+ \int_{R^n} (g(x) - \chi_E(x))\, d\mu(x) + \int_{R^n} \phi(x)p_\delta(x)\, d\mu(x). \qquad (3)$$

We estimate the terms

$$\left| \int_{R^n} (g(x) - \chi_E(x))\, d\mu(x) \right| \leq |\mu|(\{x \in Q : g(x) \neq \chi_E(x)\}) \leq \delta,$$

and

$$\left| \int_{R^n} \phi(x)p_\delta(x)\, d\mu(x) \right| \leq \|\phi p_\delta\|_{L^\infty} |\mu|(Q) \leq c_1\delta.$$

Thus, from (3) we see that

$$\nu(E) \leq c\delta + c\sigma(c\delta^{1/q}) \leq \varepsilon, \quad \text{for } \delta \text{ small enough}.$$

Whence, the restriction of μ to the cube Q is absolutely continuous, and the proof is complete. ∎

We consider now the case $q = \infty$. We want to show that it is possible to represent a linear functional $\ell \in (H^p_w(R^n))^*$ by means of a locally integrable function g in $\mathcal{A}^*$, extended from $H^p_w(R^n) \cap \mathcal{A}$ to $H^p_w(R^n) \cap \tilde{\mathcal{A}}$; this can be achieved by an approximation argument.

Given $g \in \mathcal{A}^*$ and $f \in H^p_w(R^n) \cap \tilde{\mathcal{A}}$, we want to find a sequence $\{f_j\} \subset H^p_w(R^n) \cap \tilde{\mathcal{A}}$ such that $\|f_j - f\|_{H^p_w} \to 0$ and $\int_{R^n} (f_j(x) - f(x))g(x)\, dx \to 0$ as $j \to \infty$; note that fg is integrable when $f \in \tilde{\mathcal{A}}$ and $g \in \mathcal{A}^*$. First, if f has compact support contained in $\{|x| \leq R\}$, we can approximate it by continuous functions f_j, say, also supported in $\{|x| \leq R\}$, with vanishing moments up to order N, the same as f, and such that $|\{x : |f(x) - f_j(x)| > 2^{-j}\}| \leq 2^{-j}$. It then follows that $\int_{R^n} (f_j(x) - f(x))\, dx \to 0$ and also, by Lemma 4, that $\|f_j - f\|_{H^p_w} \to 0$ as $j \to \infty$.

Finally, we can approximate a general function $f \in H_w^p(R^n) \cap \tilde{A}$ by compactly supported functions f_j in $H_w^p(R^n) \cap \tilde{A}$ such that $f(x) = f_j(x)$ for $|x| \leq 2^j$, and

$$\|f_j(1 + |\cdot|)^{t_1}\|_{L^\infty} \leq c\|f(1 + |\cdot|)^{t_1}\|_{L^\infty}, \quad \text{in case 1},$$

and

$$\|f_j(1 + |\cdot|)^{t_0}\|_{L^\infty} \leq c\|f(1 + |\cdot|)^{t_0}\|_{L^\infty},$$

and

$$\|(f - f_j)\varrho(|\cdot|)(1 + |\cdot|)^{t_1 - n}\|_{L^1} \leq 2^{-j},$$

in case 2.

Thus, $\int_{R^n} (f_j(x) - f(x)) g(x)\, dx \to 0$ as $j \to \infty$. To show that $\|f_j - f\|_{H_w^p} \to 0$ as $j \to \infty$, we go back to the proof of Lemma 4 and consider both cases there. We discuss case 1 first, i.e., $t_0 \neq n + N + 1$. It readily follows, with $F_j(x, t) = f_j * \phi_t(x)$, that

$$M_1''(F - F_j, x) \leq \begin{cases} 0 & \text{if } |x| \leq c2^j \\ c|x|^{t_1}\|f\|_{\tilde{A}} & \text{if } |x| > 2^j. \end{cases}$$

From this we see that $\|M_1''(F - F_j)\|_{L_w^p} \to 0$ as $j \to \infty$. Otherwise, since

$$(f - f_j) * \phi_t(x) = 0, \quad \text{when } t < c2^j, \quad |x - y| < t \quad \text{and } (1 + |x|)/5 \leq t,$$

we estimate

$$M_1'(F - F_j, x) \leq c(2^j + |x|)^{-t_2}\|f\|_{\tilde{A}}.$$

It thus follows that $\|M_1'(F - F_j)\|_{L_w^p} \to 0$ as $j \to \infty$. This completes the proof in case 1; case 2 is handled in a similar way.

We can now extend the representation of a linear functional $\ell \in (H_w^p(R^n))^*$ by means of a function g in $\mathcal{A}^*$ from $H_w^p(R^n) \cap \mathcal{A}$ to $H_w^p(R^n) \cap \tilde{\mathcal{A}}$. Indeed, suppose $f \in H_w^p(R^n) \cap \tilde{\mathcal{A}}$ is approximated by a sequence $\{f_j\} \subset H_w^p(R^n) \cap \mathcal{A}$. Then,

$$\ell(f) = \lim_{j \to \infty} \ell(f_j) = \lim_{j \to \infty} \int_{R^n} f_j(x) g(x)\, dx = \int_{R^n} f(x) g(x)\, dx.$$

By the basic inequality, Proposition 1 in Chapter IX, if $f \in \hat{\mathcal{D}}_0$ and $M_{1,\nu}^{\sharp, \mathcal{P}_m} g \in L_w^{p'}(R^n)$, it follows that

$$\left| \int_{R^n} f(x) g(x)\, dx \right| \leq c\|f\|_{H_w^p} \|M_{1,\nu}^{\sharp, \mathcal{P}_m} g\|_{L^{p'}{}_w}, \quad 1 \leq p < \infty.$$

We note in passing that Theorem 2 in Chapter IX gives the corresponding result for $0 < p < 1$. Now, from the Hahn-Banach theorem, it then follows that the mapping $f \rightarrow \int_{R^n} f(x)g(x)\,dx$ extends to a continuous linear functional ℓ on $H^p_w(R^n)$ such that

$$\ell(f) = \int_{R^n} f(x)g(x)\,dx\,, \quad f \in \hat{\mathcal{D}}_0\,, \quad \|\ell\| \le c\|M^{\sharp,\mathcal{P}_m}_{1,\nu}g\|_{L^{p'}_w}\,.$$

As discussed above, there is a g in $\mathcal{A}^*$ such that

$$\ell(f) = \int_{R^n} f(x)g_1(x)\,dx\,, \quad f \in H^p_w(R^n) \cap \mathcal{A} \quad (\text{or } H^p_w(R^n) \cap \tilde{\mathcal{A}})\,.$$

In particular, if $f \in \hat{\mathcal{D}}_0$, we get

$$\int_{R^n} f(x)g_1(x)\,dx = \int_{R^n} f(x)g(x)\,dx\,,$$

and from this we conclude that $g - g_1$ is a polynomial. If also $g \in \mathcal{A}^*$, then $g - g_1$ is a polynomial of degree less than or equal to N. Thus, since functions $f \in H^p_w(R^n) \cap \mathcal{A}$ (or $H^p_w(R^n) \cap \tilde{\mathcal{A}}$), have vanishing moments of order less than or equal to N, we get that for these functions, $\ell(f) = \int_{R^n} f(x)g(x)\,dx$.

On the other hand, if $\ell \in (H^p_w(R^n))^*$, as we have seen, there is a function $g \in \mathcal{A}^*$ such that

$$\ell(f) = \int_{R^n} f(x)g(x)\,dx\,, \quad \text{all} \quad f \in H^p_w(R^n) \cap \mathcal{A} \quad (\text{or } H^p_w(R^n) \cap \tilde{\mathcal{A}})\,.$$

To complete the proof we want to prove that

$$\|M^{\sharp,\mathcal{P}_m}_{q',\nu}g\|_{L^{p'}_w} \le \|\ell\|\,, \quad m \text{ large enough.}$$

To see this, let $\{Q_j\}_{j=1}^\infty$ be the countable family consisting of all cubes in R^n whose center has rational coordinates and whose sides have rational length; in the definition of the sharp maximal function it suffices to take the supremum over the cubes $\{Q_j\}$. Moreover, by the Monotone Convergence Theorem, it is enough to take the supremum over the family $\{Q_j\}_{j=1}^T$, T large, and estimate the $L^{p'}_w(R^n)$-norm of $M^{\sharp,\mathcal{P}_m}_{q',\nu}g$ uniformly in T. Note that, in the process, we set $M^{\sharp,\mathcal{P}_m}_{q',\nu}g(x) = 0$ if $x \notin \bigcup_{j=1}^T Q_j$.

Let

$$S_j = \frac{|Q_j|}{\nu(Q_j)}\left(\frac{1}{|Q_j|}\int_{Q_j}|f(x) - p_j(x)|^{q'}\,dx\right)^{1/q'}\,, \quad 1 \le j \le T\,,$$

where p_j is the unique polynomial of degree less than or equal to m such that

$$\int_{Q_j} f(x)x^\alpha\,dx = \int_{Q_j} p_j(x)x^\alpha\,dx\,, \quad |\alpha| \le N\,.$$

We want to show that

$$\left\| \max_{1\le j\le T} S_j\chi_{Q_j} \right\|_{L^{p'}_w} \le c\|\ell\|\,.$$

Let $\{E_j\}_{j=1}^T$ be a family of pairwise disjoint sets such that $E_j \subseteq Q_j$, $\bigcup_{j=1}^T E_j = \bigcup_{j=1}^T Q_j$, and so that

$$S_j = \max_{1\le i\le T} S_i\chi_{Q_i}(x)\,, \quad x\in E_j\,.$$

It suffices to show that

$$\left\| \sum_{j=1}^T S_j\chi_{E_j} \right\|_{L^{p'}_w} \le c\|\ell\|\,.$$

Let

$$b_j(x) = |g(x) - p_j(x)|^{q'-2}(g(x) - p_j(x))\chi_{Q_j}(x)\,.$$

Note that $\|b_j\|_{L^\infty} \le 1$ if $q = \infty$, and

$$\|b_j\|_{L^q} = \left(\int_{Q_j} |g(x) - p_j(x)|^{q'}\,dx\right)^{1/q} = (S_j\nu(Q_j)/|Q_j|)^{q'-1}|Q_j|^{1/q}\,,$$

if $1 < q < \infty$.

Let $\tilde{p}_j \in \mathcal{P}_m$ be the unique polynomial such that

$$\int_{Q_j} b_j(x)x^\alpha\,dx = \int_{Q_j} \tilde{p}_jx^\alpha\,dx\,, \quad |\alpha| \le m\,;$$

and note that $\|\tilde{p}_j\|_{L^q} \le c\|b_j\|_{L^q}$. Also observe that

$$\int_{Q_j} (b_j(x) - \tilde{p}_j(x))g(x)\,dx = \int_{Q_j} (b_j(x) - \tilde{p}_j(x))(g(x) - p_j(x))\,dx$$

$$= \int_{Q_j} b_j(x)(g(x) - p_j(x))\,dx = \int_{Q_j} |g(x) - p_j(x)|^{q'}\,dx$$

$$= (S_j\nu(Q_j)/|Q_j|)^{q'}|Q_j|\,.$$

Now, if b_j does not vanish identically we put

$$a_j(x) = (|Q_j|^{1-1/q'}/\|b_j\|_{L^q})(b_j(x) - \tilde{p}_j(x))\chi_{Q_j}(x)\,.$$

Then, since $\int_{R^n} a_j(x)g(x)\,dx = S_j\nu(Q_j)$, with the normalization

$$\|a_j\|_{L^q} \leq |Q_j|^{1-1/q'}(1 + \|\tilde{p}_j \chi_{Q_j}\|_{L^q}/\|b_j\|_{L^q}) \leq c|Q_j|^{1/q},$$

$a_j(x)$ is an atom with support contained in Q_j and with vanishing moments up to order $m-1$.

If $p = 1$, and hence $p' = \infty$, note that

$$|\ell(a_j)| \leq \|\ell\|\,\|a_j\|_{H_w^1} \leq c\|\ell\|\,\|\chi_{Q_j}\|_{L^1} = c\|\ell\|\,|Q_j|.$$

Since $\ell(a_j) = \int_{R^n} a_j(x)g(x)\,dx$, it follows that

$$S_j \leq S_j\nu(Q_j)/\nu(E_j) \leq c\|\ell\|,$$

and consequently, $\|\sum_{j=1}^T S_j\chi_{E_j}\|_{L^\infty} \leq c\|\ell\|$.

On the other hand, if $1 < p < \infty$, we set

$$\lambda_j = S_j^{p'/p}\nu(E_j)/\nu(Q_j).$$

With this choice for the λ_j's we have

$$I = \ell\Big(\sum_{j=1}^T \lambda_j a_j\Big) = \int_{R^n}\sum_{j=1}^T \lambda_j a_j(x)g(x)\,dx$$

$$= \sum_{j=1}^T \lambda_j\nu(Q_j) = \sum_{j=1}^T S_j^{p'}\nu(E_j) = \left\|\sum_{j=1}^T S_j\chi_{E_j}\right\|_{L_w^{p'}}.$$

Moreover, also

$$I \leq \|\ell\|\left\|\sum_{j=1}^T \lambda_j a_j\right\|_{H_w^p} \leq c\|\ell\|\left\|\sum_{j=1}^T \lambda_j\chi_{Q_j}\right\|_{L_w^p}.$$

We estimate the $L_w^p(R^n)$-norm above by duality; it equals

$$\sup_{\|h\|_{L_w^{p'}} \leq 1}\int_{R^n}\sum_{j=1}^T \lambda_j\chi_{Q_j}(x)h(x)w(x)\,dx.$$

Now, for each $h(x)$ the above integral is

$$\sum_{j=1}^T \lambda_j\int_{Q_j} h(x)w(x)\,dx = \sum_{j=1}^T S_j^{p'/p}\nu(E_j)\frac{1}{\nu(Q_j)}\int_{Q_j} h(x)w(x)\,dx$$

$$\leq \int_{R^n}\sum_{j=1}^T S_j^{p'/p}\chi_{E_j}(y)M_\nu h(y)w(y)\,dy.$$

Whence, since M_ν is bounded on $L_w^{p'}(R^n)$, $1 < p' \leq \infty$, and $\|h\|_{L_w^{p'}} \leq 1$, the above expression is dominated by

$$c \left\| \sum\nolimits_{j=1}^{T} S_j^{p'/p} \chi_{E_j} \right\|_{L_w^p} \|h\|_{L_w^{p'}} \leq c \left\| \sum\nolimits_{j=1}^{T} S_j \chi_{E_j} \right\|_{L_w^{p'}}^{p'/p}.$$

Thus, combining the above estimates we get

$$\left\| \sum\nolimits_{j=1}^{T} \lambda_j \chi_{E_j} \right\|_{L_w^{p'}}^{p'} \leq c\|\ell\| \left\| \sum\nolimits_{j=1}^{T} S_j \chi_{E_j} \right\|_{L_w^{p'}}^{p'/p},$$

and we conclude that

$$\left\| \sum\nolimits_{j=1}^{T} \lambda_j \chi_{E_j} \right\|_{L_w^{p'}} \leq c\|\ell\|, \quad \text{and} \quad \left\| M_{q',\nu}^{\natural, \mathcal{P}_m} g \right\|_{L_w^{p'}} \leq c\|\ell\|.$$

This completes the proof. ∎

As indicated above, the proof for $0 < p < 1$ follows along similar lines. Note that the smallest value for the integer $m = m(p,w)$ for which the argument holds is the same value as the smallest order of the moment condition $\int_{R^n} a_j(x)x^\alpha\, dx = 0$, $|\alpha| \leq m$, so that the inequality $\| \sum_j \lambda_j a_j \|_{H_w^p} \leq c\| \sum_j \lambda_j \chi_{Q_j} \|_{L_w^p}$ is true, cf. Chapter VIII.

Sources and Remarks. That the dual of $H^1(R^n)$ is $BMO(R^n)$ is a celebrated result of C. Fefferman [1971], and it is the point of departure of a great deal of the theory of Hardy spaces of several real variables. The dual of the weighted spaces $H_w^p(R^n)$, $0 < p \leq 1$ was described by J. García-Cuerva [1979], see also E. Lotkowski and R. Wheeden [1976]. García-Cuerva's results extend the work of P. Duren, B. Romberg and A. Shields [1969] concerning the dual of $H^p(T)$ as well. Equivalent expressions for the norms involved in the description of $(H_w^p(R^n))^*$ have been given in Chapter III.

XI

Singular Integrals and Multipliers

A bounded function $m(\xi)$ defined in $R^n \setminus \{0\}$ is called a multiplier. A multiplier operator T is defined by the expression

$$(Tf)\hat{\ }(\xi) = m(\xi)\hat{f}(\xi), \quad f \in \hat{\mathcal{D}}_0 .$$

We say that T is a bounded multiplier mapping on $H^p_w(R^n)$ with norm less than or equal to c, if

$$\|Tf\|_{H^p_w} \le c\|f\|_{H^p_w} . \tag{1}$$

In this case T admits a continuous extension as a linear mapping on $H^p_w(R^n)$ with norm not exceeding c. If the function $m(\xi)$ is in addition homogeneous of degree 0, we say that T is a Calderón-Zygmund singular integral operator. A basic result of R. Hunt, B. Muckenhoupt and R. Wheeden [1973] states that when $n = 1$ and $1 < p < \infty$, a singular integral operator, that is, the Hilbert transform, is bounded in $L^p_w(R^n)$ if and only if $w \in A_p$. Thus, we will be mostly interested in estimates such as (1) for either $0 < p \le 1$, or $p > 1$ and $w \notin A_p$. Nevertheless, the results proved in this chapter include Hörmander's multiplier theorem for $L^p_w(R^n)$ spaces. The main method of proof used in this chapter is the atomic decomposition of the weighted Hardy spaces. In addition, we present other methods, including pointwise estimates involving Lusin and Littlewood-Paley integrals, as well as pointwise estimates in terms of sharp maximal functions. In fact, we find that each method is more appropriate for a different range of values of p, depending on the weight w.

We begin by introducing a notation. By $|x| \sim t$ we denote the fact that the values of x lie in the annulus $\{x \in R^n : at < |x| < bt\}$, where $0 < a \le 1 <$

$b < \infty$ are values specified in each instance. Altough this definition requires a determination of a and b, such values are often unimportant and the choice $a = 1/2$ and $b = 2$ usually works. Unless a choice of a and b is necessary, we do not specify their values; this indetermination will create no confusion.

Let $\ell \geq 0$ be a real number and $1 \leq q \leq 2$. We say that the multiplier m satisfies the condition $M(q, \ell)$, and we often write $m \in M(q, \ell)$, if

$$\left(\int_{\{|\xi| \sim R\}} |D^\alpha m(\xi)|^q \, d\xi \right)^{1/q} \leq cR^{(n/q)-|\alpha|}$$

for all $R > 0$ and multi-indices α with $|\alpha| \leq \ell$ when ℓ is a positive integer, and, in addition,

$$\left(\int_{\{|\xi| \sim R\}} |D^\alpha m(\xi) - D^\alpha m(\xi - z)|^q \, d\xi \right)^{1/q} \leq c(|z|/R)^\gamma R^{(n/q)-|\alpha|}$$

for all $|z| < R/2$ and all multi-indices α with $|\alpha| = j =$ integer part of ℓ, and $\ell = j + \gamma$ when ℓ is not an integer.

We reserve the notation $k(x)$ for the kernel that corresponds to the inverse Fourier transform of m in the sense of distributions. It is a natural question to consider how the behaviour of k reflects the fact that $m \in M(q, \ell)$; first a definition.

For a real number $\tilde{\ell} \geq 0$ and $1 \leq \tilde{q} \leq \infty$, we say that k verifies the condition $\tilde{M}(\tilde{q}, \tilde{\ell})$, and we often write $k \in \tilde{M}(\tilde{q}, \tilde{\ell})$, if

$$\left(\int_{\{|x| \sim R\}} |D^\beta k(x)|^{\tilde{q}} \, dx \right)^{1/\tilde{q}} \leq cR^{(n/\tilde{q})-n-|\beta|}, \quad \text{all } R > 0,$$

for all multi-indices $|\beta| < \tilde{\ell}$, and, in addition, if $\tilde{j}$ denotes the largest integer strictly less than $\tilde{\ell}$ and $\tilde{\ell} = \tilde{j} + \tilde{\gamma}$,

$$\left(\int_{\{|x| \sim R\}} |D^\beta k(x) - D^\beta k(x - z)|^{\tilde{q}} \right)^{1/\tilde{q}}$$
$$\leq \begin{cases} c(|z|/R)^{\tilde{\gamma}} R^{(n/\tilde{q})-n-\tilde{j}} & \text{if } 0 < \tilde{\gamma} < 1 \\ c(|z|/R)(\ln(|z|/R)) R^{(n/\tilde{q})-n-\tilde{j}} & \text{if } \tilde{\gamma} = 1, \end{cases}$$

for all $|z| < R/2$, $R > 0$, and all multi-indices β with $|\beta| = \tilde{j}$. In case $\tilde{q} = \infty$, the integral expression above should be replaced by a supremum norm in the usual way.

We open with some remarks. At a first glance, the condition $\tilde{M}(\tilde{q}, \tilde{\ell})$ may appear a bit unusual when $\tilde{\ell}$ is an integer. It may seem more natural to use all multi-indices β with $|\beta| \leq \tilde{\ell}$ in the first expression above rather than considering the differences. However, such condition is too restrictive as our next result shows.

Lemma 1. Suppose $m \in M(q, \ell)$, $1 \le q \le 2$. Given $1 \le \tilde{q} \le \infty$, let $p \ge 1$ be such that $1/p = \max(1/q, 1 - 1/\tilde{q})$. Then $k \in \tilde{M}(\tilde{q}, \tilde{\ell})$, where $\tilde{\ell} = \ell - n/p$.

Observation. We have assumed that the multiplier m is a bounded function. However, the condition $M(q, \ell)$ on m already implies that m is bounded provided that $\ell > n/q$. Moreover, the functions $m(\xi/R)$ have Lipschitz $\ell - n/q$ norms which are uniformly bounded in R in the annulus $\{|\xi| \sim 1\}$ if $0 < \ell - n/q < 1$; to prove this one may use Sobolev's inequality on expressions of BMO-type, the details are left to the reader.

Now, let the "$M(q, \ell)$ norm of m" denote the infimum of all possible constants c in the definition of the condition $M(q, \ell)$; similarly we may define the "$\tilde{M}(\tilde{q}, \tilde{\ell})$ norm of the kernel k". Also, if η is an arbitrary Schwartz function, then multiplication of either m or k by η only increases the $M(q, \ell)$ or the $\tilde{M}(\tilde{q}, \tilde{\ell})$ norms by at most a constant factor c_η, depending solely on η. This is readily seen by using Leibniz's differentiation rule, and it is especially simple when η is supported in an annulus. We also note that the condition $M(q, \ell)$ is invariant under the dilation $m(\xi) \to m(t\xi)$, $t > 0$, and similarly, the condition $\tilde{M}(\tilde{q}, \tilde{\ell})$ is invariant under the dilation $k(x) \to t^{-n} k(x/t)$, $t > 0$. Therefore, if $m \in M(q, \ell)$, then so does $\eta(t\xi) m(\xi)$, uniformly in $t > 0$. Similarly, if $k \in \tilde{M}(\tilde{q}, \tilde{\ell})$, also does $\eta(tx) k(x)$, uniformly in $t > 0$.

Finally we observe that the conditions $M(q, \ell)$ and $\tilde{M}(\tilde{q}, \tilde{\ell})$ are monotonic in q, ℓ and $\tilde{q}, \tilde{\ell}$, respectively. In other words, if $q \ge q_1$ and $\ell \ge \ell_1$, then $M(q, \ell)$ implies $M(q_1, \ell_1)$, and similarly, if $\tilde{q} \ge \tilde{q}_1$ and $\tilde{\ell} \ge \tilde{\ell}_1$, then $\tilde{M}(\tilde{q}, \tilde{\ell})$ implies $\tilde{M}(\tilde{q}_1, \tilde{\ell}_1)$. This observation follows from Hölder's inequality and the mean value theorem together with Minkowski's inequality.

Proof of Lemma 1. Since $q \ge p$ and $\tilde{q} \le p'$, by the monotonicity in the conditions, $M(q, \ell)$ implies $M(p, \ell)$ and $\tilde{M}(p', \tilde{\ell})$ implies $\tilde{M}(\tilde{q}, \tilde{\ell})$. Hence, it suffices to show that if $\tilde{\ell} = \ell - n/p$ and $m \in M(p, \ell)$, then $k \in \tilde{M}(p', \tilde{\ell})$.

There are several cases depending on whether ℓ is an integer and on whether we are estimating expressions involving $D^\beta k(x)$ or the differences $D^\beta k(x) - D^\beta k(x - z)$. First we split the multiplier m into several parts, and treat each one of them separately; the proof is then based on the Haussdorff-Young inequality. We do not do all cases, but we hope that the indications we give in the cases we discuss in detail are sufficient to make it clear how the remaining cases are handled.

First, since $k(x) \to t^{-n} k(x/t)$ corresponds on the Fourier transform side to $m(\xi) \to m(t\xi)$, $t > 0$, and the conditions $\tilde{M}(\tilde{q}, \tilde{\ell})$ and $M(q, \ell)$ are invariant under these dilations, we may assume that $R = 1$. Let ϕ be a nonnegative Schwartz function with support contained in $\{1/2 < |y| < 2\}$, such that $\sum_{i=-\infty}^{\infty} \phi(2^{-i} y) = 1$ for $y \ne 0$. Set $\eta(\xi) = 1 - \sum_{i=1}^{\infty} \phi(2^{-i}\xi)$ and

$$m_0(\xi) = \eta(\xi) m(\xi), \quad m_i(\xi) = \phi(2^{-i}\xi) m(\xi), \quad i = 1, 2, \ldots$$

and let $k_i(x)$ denote the inverse Fouries transform of $m_i(\xi)$, $i = 0, 1, 2, \ldots$

Case 1. We start, when $i = 0$, by estimating the expressions $D^\beta k_0(x)$ and $D^\beta k_0(x) - D^\beta k_0(x - z)$, which are essentially the inverse Fourier transforms of $\xi^\beta \eta(\xi) m(\xi)$ and $(1 - e^{iz \cdot \xi}) \eta(\xi) m(\xi)$, respectively. Since $|\eta(\xi)| \leq 1$ and η has support contained in $\{|\xi| < 2\}$, we get

$$
\left(\int_{\{|x| \sim 1\}} |D^\beta k_0(x)|^{p'} dx \right)^{1/p'} \leq c \left(\int_{R^n} |\xi^\beta \eta(\xi) m(\xi)|^p d\xi \right)^{1/p}
$$

$$
\leq \left(\sum_{i=0}^\infty \int_{\{2^{-i} < |\xi| < 2^{-i+1}\}} |\xi|^{|\beta| p} |m(\xi)|^p d\xi \right)^{1/p}
$$

$$
\leq c \sum_{i=0}^\infty 2^{-i|\beta|} \left(\int_{\{|\xi| \sim 2^{-i}\}} |m(\xi)|^p d\xi \right)^{1/p}
$$

$$
\leq c \sum_{i=0}^\infty 2^{-i(|\beta| + n/p)} \leq c.
$$

Now, when estimating $D^\beta k_0(x) - D^\beta k_0(z - x)$, we have in addition the factor $1 - e^{iz \cdot \xi}$ on the Fourier transform side, and since $|1 - e^{iz \cdot \xi}| \leq |z| |\xi|$, when $|\xi| \leq 2$ we get

$$
\left(\int_{\{|x| \sim 1\}} |D^\beta k_0(x) - D^\beta k_0(x - z)|^{p'} dx \right)^{1/p'} \leq c|z|.
$$

Case 2. Next we consider $D^\beta k_i(x)$, $i > 0$. Since $\sum_{|\alpha|=j} |x^\alpha| \geq c > 0$ for $|x| \sim 1$, when ℓ is an integer we get

$$
\left(\int_{\{|x| \sim 1\}} |D^\beta k_i(x)|^{p'} dx \right)^{1/p'} \leq c \left(\int_{\{|x| \sim 1\}} \left(\sum_{|\alpha|=j} |x^\alpha D^\beta k_i(x)| \right)^{p'} dx \right)^{1/p'}
$$

$$
\leq c \sum_{|\alpha|=j} \left(\int_{\{|x| \sim 1\}} |x^\alpha D^\beta k_i(x)|^{p'} dx \right)^{1/p'}.
$$

By the Hausdorff-Young inequality this expression does not exceed

$$
c \sum_{|\alpha|=j} \left(\int_{R^n} |D^\alpha(\xi^\beta m_i(\xi))|^p d\xi \right)^{1/p}.
$$

Now, since $\xi^\beta \phi(\xi)$ is a Schwartz function we see that $2^{-i|\beta|} \xi^\beta m_i(\xi) = (2^{-i}\xi)^\beta \phi(2^{-i}\xi) m(\xi)$ satisfies the condition $M(p, \ell)$ uniformly in i, and furthermore, its support lies in $\{|\xi| \sim 2^i\}$. Hence, for each α in the above sum, the terms are bounded by $c2^{i(n/p+|\beta|-|\alpha|)}$ and, since $|\alpha| = j = \ell$ when ℓ is an integer, we conclude that

$$\left(\int_{\{|x|\sim 1\}} |D^\beta k_i(x)|^{p'}\, dx \right)^{1/p'} \le c2^{i(n/p+|\beta|-\ell)}. \tag{2}$$

If ℓ is not an integer, i.e., $\ell = j + \gamma$ with $0 < \gamma < 1$, we use the inequality

$$\sum_{|\alpha|=j} |x^\alpha \sin(x \cdot z_\alpha)| \ge c > 0 \quad \text{for } |x| \sim 1,$$

for some suitably chosen z_α; in fact we may choose $z_\alpha = \alpha/|\alpha|$. Since multiplication by the sine factors corresponds to taking differences in the Fourier transform side, as before we see that

$$\left(\int_{\{|x|\sim 1\}} |D^\beta k_i(x)|^{p'}\, dx \right)^{1/p'}$$

$$\le c \sum_{|\alpha|=j} \left(\int_{\{|x|\sim 1\}} |x^\alpha \sin(x \cdot z_\alpha) D^\beta k_i(x)|^{p'}\, dx \right)^{1/p'}$$

$$\le c \sum_{|\alpha|=j} \left(\int_{R^n} |D^\alpha((\xi + z_\alpha)^\beta m_i(\xi + z_\alpha)) - D^\alpha((\xi - z_\alpha)^\beta m_i(\xi - z_\alpha))|^p\, d\xi \right)^{1/p}.$$

Since $2^{-i|\beta|}\xi^\beta m_i(\xi) \in M(p,\ell)$, $\ell = j + \gamma$, its support is contained in the annulus $\{|\xi| \sim 2^i\}$ and $| \pm z_\alpha| \le 1 < 2^i/2$, each term in the last sum above is bounded by $c(|z_\alpha|/2^i)^\gamma 2^{i(n/p+|\beta|-j)} \le c2^{(n/p+|\beta|-\ell)}$. Thus, we also get (2) when ℓ is not an integer.

Case 3. Finally we estimate the terms involving the difference $D^\beta k_i(x) - D_\beta k_i(x - z)$ when $|\beta| = \tilde{j}$, where $\tilde{j}$ is the largest integer less than $\tilde{\ell}$. When ℓ is an integer we first invoke the inequality $\sum_{|\alpha|=\ell} |x^\alpha| \ge c > 0$, and obtain

$$\left(\int_{\{|x|\sim 1\}} |D^\beta k_i(x) - D^\beta k_i(x - z)|^{p'}\, dx \right)^{1/p'}$$

$$\le c \sum_{|\alpha|=\ell} \left(\int_{\{|x|\sim 1\}} |x^\alpha(D^\beta k_i(x) - D^\beta k_i(x - z))|^{p'}\, dx \right)^{1/p'}.$$

The Fourier transform of the function $x^\alpha(D^\beta k_i(x) - D^\beta k_i(x - z))$ is essentially equal to $D^\beta((1 - e^{i\xi \cdot z})\xi^\beta m_i(\xi))$, which, by Leibnitz's differentiation formula, can be expressed as a sum of products of terms of the form

$$c_{\alpha_1} D^{\alpha_1}(1 - e^{i\xi \cdot z}) D^{\alpha-\alpha_1}(\xi m_i(\xi)),$$

with all multi-indices α_1 with $0 \leq \alpha_1 \leq \alpha$. We now use the estimates $|1 - e^{i\xi \cdot z}| \leq |\xi||z|$ and $|D_1^\alpha(1 - e^{iz \cdot x})| \leq c|z|^{|\alpha_1|}$ for $\alpha_1 \neq 0$ and $|x| \sim 1$, together with the condition $M(p, \ell)$ which is satisfied uniformly by the functions $2^{-i|\beta|}\xi^\beta m_i(\xi)$. Thus, by the Hausdorff-Young inequality we get

$$\left(\int_{\{|x| \sim 1\}} |x^\alpha (D^\beta k_i(x) - D^\beta k_i(x - z))|^{p'} \, dx \right)^{1/p'}$$

$$\leq c2^{i(n/p+|\beta|-|\alpha|)}|z|2^i + c \sum_{0 < \alpha_1 \leq \alpha} 2^{i(n/p+|\beta|-|\alpha-\alpha_1|)}|z|^{|\alpha-1|}$$

$$\leq c2^{i(n/p+|\beta|+1-|\alpha|)}|z|,$$

when $|z| \leq 1$. Thus, when ℓ is an integer,

$$\left(\int_{\{|x| \sim 1\}} |D^\beta k_i(x) - D^\beta k_i(x - z)|^{p'} \, dx \right)^{1/p'} \leq c2^{i(n/p+|\beta|-\ell)}|z|2^i.$$

Note that when $2^i > 1/|z|$ we get a better estimate using the triangle inequality and the estimate in case 2.

When ℓ is not an integer we may use sine factors instead, as we did in case 2, and note that this corresponds to taking differences in the Fourier transform side. After using Leibniz's formula we get, in the Fourier transform side, a sum involving expressions of the form

$$c_{\alpha_1} D^{\alpha_1}(e^{i(\xi - z_\alpha) \cdot z} - e^{i\xi \cdot z})D^{\alpha - \alpha_1}\xi^\beta m_i(\xi)$$
$$+ c_{\alpha_1} D^{\alpha_1}(1 - e^{i\xi \cdot z})D^{\alpha - \alpha_1}(\xi^\beta m_i(\xi) - (\xi - z_\alpha)^\beta m_i(\xi - z_\alpha)),$$

for all $\alpha_1 \leq \alpha$ and all $|\alpha| = j$, where z_α is defined as in case 2, and $\ell = j + \gamma$.

In addition to the estimates obtained in case ℓ is an integer, we need the estimates

$$\left| D^{\alpha_1}(e^{i(\xi - z_\alpha) \cdot z} - e^{i\xi \cdot z}) \right| \leq |z_\alpha||z|^{|\alpha_1|+1} \quad 0 \leq |\alpha_1| \leq j,$$

and the fact that the multipliers $2^{-i|\beta|}\xi^\beta m_i(\xi)$ satisfy the condition $M(q, \ell - |\alpha_1|)$ uniformly in i for $0 \leq |\alpha_1| \leq j$. We thus obtain

$$\left(\int_{\{|x| \sim 1\}} |D^\beta k_i(x) - D^\beta k_i(x - z)|^{p'} \, dx \right)^{1/p'}$$

$$\leq c2^{i(n/p+|\beta|-j)}|z| + c2^{i(n/p+|\beta|-j-\gamma)}|z|2^i$$

$$+ \sum_{d=1}^{j}(2^{i(n/p+|\beta|-j+d)})|z|^{d+1} + 2^{i(n/p+|\beta|-j+d-\gamma)}|z|^d$$

$$\leq c2^{i(n/p+|\beta|-\ell)}|z|2^i,$$

provided $2^i < 1/|z|$. When $2^i \geq 1/|z|$, we use instead the triangle inequality and the estimates of case 2.

By summing the estimates in case 1 and case 2 it follows that

$$
\left(\int_{\{|x|\sim 1\}} |D^\beta k(x)|^{p'} \, dx \right)^{1/p'}
$$

$$
\leq \left(\int_{\{|x|\sim 1\}} |D^\beta k_0(x)|^{p'} \, dx \right)^{1/p'} + \sum_{i=1}^\infty \left(\int_{\{|x|\sim 1\}} |D^\beta k_i(x)|^{p'} \, dx \right)^{1/p'}
$$

$$
\leq c + c \sum_{i=1}^\infty 2^{i(n/p+|\beta|-\ell)} \leq c
$$

provided $|\beta| < \tilde{\ell} = \ell - n/p$.

Also, by summing the estimates obtained in cases 1, 2 and 3, we get, for $|\beta| = \tilde{\jmath}$,

$$
A = \left(\int_{\{|x|\sim 1\}} |D^\beta k(x) - D^\beta k(x - z)|^{p'} \, dx \right)^{1/p'}
$$

$$
\leq I_0 + \sum_{i,\, 1 < 2^i < 1/|z|} I_i + \sum_{i,\, 2^i \geq 1/|z|} I_i \,,
$$

where

$$
I_i = \left(\int_{\{|x|\sim 1\}} |D^\beta k_i(x) - D^\beta k_i(x - z)|^{p'} \, dx \right)^{1/p'}, \quad i = 0, 1, 2, \ldots
$$

Thus,

$$
A \leq c|z| + c \sum_{i,\, 1 < 2^i < 1/|z|} |z| 2^{i(n/p+|\beta|-\ell+1)} + c \sum_{i,\, 2^i \geq 1/|z|} 2^{i(n/p+|\beta|-\ell)}
$$

$$
\leq c|z| + c \sum_{i,\, 1 < 2^i < |z|} 2^{i(1-\tilde{\gamma})} + c \sum_{i,\, 2^i \geq 1/|z|} 2^{i\tilde{\gamma}}
$$

$$
\leq \begin{cases} c|z|^{\tilde{\gamma}} & \text{for } |z| < 1 \text{ when } 0 < \tilde{\gamma} < 1 \\ c|z| \ln(2/|z|) & \text{for } |z| < 1 \text{ when } \tilde{\gamma} = 1. \end{cases}
$$

This completes the proof. ∎

We pass now to consider the action of the kernels on atoms.

Lemma 2. Let $1 \leq q \leq 2$, and suppose that the multiplier $m \in M(q, \ell)$. Also, let N_0 be a nonnegative integer and let $1 \leq p_0 \leq \infty$. Further, if a is a (p_0, N_0) atom normalized so that its support is contained in $B(x_0, r)$, and ϕ is the function introduced in Lemma 1, let $\phi_i(x) = \phi(2^{-i-2}x/r)$, and $k_i(x) = \phi_i(x)k(x)$, $i = 1, 2, \ldots$ Also let $1/p_1 = \max(1/q, 1 - 1/p_0)$ and $\tilde{\ell} = \ell - n/p_1$. Finally, for $i = 1, 2, 3$, set

$$
\tilde{c}_i = \begin{cases}
c2^{i(n+N_0+1)} & \text{if } \tilde{\ell} > N_0 + 1 \\
c2^{i(n+\tilde{\ell})}/i & \text{if } \tilde{\ell} \leq N_0 + 1, \tilde{\ell} \text{ integer} \\
c2^{i(n+\tilde{\ell})} & \text{if } \tilde{\ell} \leq N_0 + 1, \tilde{\ell} \text{ noninteger.}
\end{cases}
$$

Here c is a constant independent of i, chosen sufficiently large. Then, if a is a (p_0, N_0) atom with support contained in $B(x_0, r)$, $b_i = \tilde{c}_i(a * k_i)$ is a (p_0, N_0) atom with support contained in $B(x_0, 2^{i+4}r)$.

Proof. By Lemma 1 the kernel k, and consequently also the k_i's, satisfy the condition $\tilde{M}(p_0, \tilde{\ell})$, with $\tilde{\ell} = \ell - n/p_1$, uniformly in i and r. It is readily seen that the support of $a * k_i \subseteq B(x_0, 2^{i+4}r)$, and that the moments of $a * k_i$ up to order N_0 vanish. It only remains to show that for the appropriate choice of $\tilde{c}_i$, we have $\|b_i\|_{L^p} \leq |B(x_0, 2^{i+4}r)|$.

Let N be the largest integer less than or equal to N_0 such that $N < \tilde{\ell}$. Further, let $R_i(x, y)$ denote the remainder in the Taylor expansion of $k_i(x_0 - y)$, as a function of y, at the point x_0, i.e.,

$$
R_i(x, y) = \sum_{|\beta|=N} c_\beta (y - x_0)^\beta \int_0^1 (1-s)^{N-1} D^\beta k_i(x - x_0 - s(y - x_0)) \, ds .
$$

Also note that, as a function of y, $k_i(x_0 - y) - R_i(x, y)$ is a polynomial of degree at most $N - 1$.

Now, by the moment conditions on a, it follows that $a * k_i(x)$ is equal to

$$
\int_{R^n} R_i(x, y)a(y) \, dy
$$

$$
= \sum_{|\beta|=N} c_\beta \int_0^1 (1-s)^{N-1}
$$

$$
\times \left(\int_{R^n} a(y)(y - x_0)^\beta (D^\beta k_i((x - x_0) - s(y - x_0)) - D^\beta k_i(x - x_0)) \, dy \right) ds .
$$

To estimate $\|a * k_i\|_{L^{p_0}}$ we apply Minkowski's inequality, so it suffices to estimate the L^{p_0}-norm of

$$
D^\beta k_i((x - x_0) - s(y - x_0)) - D^\beta k_i(x - x_0),
$$

as a function of x for each $0 \le s \le 1$ and $y \in B(x_0, r) \supseteq \operatorname{supp} a(y)$. Since $s|y - x_0| < r$ and $|x - x_0| < 2^{i+2}r$, from the condition $\tilde{M}(p_0, \tilde{\ell})$ on k_i we see that the norm in question is bounded by

$$\begin{cases} c(2^i r)^{n/p_0 - n - N}(s|y - x_0|/2^i r)^{\tilde{\ell} - N} & \text{if } \tilde{\ell} - 1 < N < \tilde{\ell} \\ c(2^i r)^{n/p_0 - n - N}(s|y - x_0|/2^i r)\ln(2^i r/s|y - x_0|) & \text{if } N = \tilde{\ell} - 1 \\ c(2^i r)^{n/p_0 - n - N}(s|y - x_0|/2^i r) & \text{if } N < \tilde{\ell} - 1. \end{cases}$$

The last estimate follows from the mean value theorem and the estimates on the derivatives of order N in the condition $\tilde{M}(p_0, \tilde{\ell})$ on k_i. Since

$$\int_{R^n} |a(y)| \, |(y - x_0)^\beta| \, dy \le \|a\|_{L^{p_0}} r^{N + n - n/p_0} \le cr^{N+n},$$

by Minkowski's inequality it follows that

$$\|a * k_i\|_{L^{p_0}} \le (c/\tilde{c}_i)(2^i r)^{n/p_0} \le (1/\tilde{c}_i)|B(x_0, 2^{i+4} r)|^{1/p_0},$$

provided, of course, that the constant in the definition of $\tilde{c}_i$ is chosen small enough. Thus, $\tilde{c}_i(a * k_i)$ is a (N_0, p_0) atom. ■

For the rest of the kernel we make use of

Lemma 3. Let $1 \le q \le 2$ and $m \in M(q, \ell)$, N_0 a nonnegative integer, and $1 \le p_0 \le \infty$. Further assume that the multiplier operator T corresponding to m is bounded on $L^{p_0}(R^n)$. Given a (p_0, N_0) atom a with support contained in $B(x_0, r)$, let ϕ_i, $i = 1, 2, \dots$, be defined as in Lemma 2, and set $\phi_0 = 1 - \sum_{i=1}^\infty \phi_i$, and $k_0(x) = \phi_0(x)k(x)$. Also, let $\tilde{\ell} > 0$ be chosen as in Lemma 2. Then, there is a constant $\tilde{c}_0$, independent of r, such that if $b_0 = \tilde{c}_0(a * k_0)$, b_0 is a (p_0, N_0) atom with support contained in $B(x_0, 2^4 r)$.

Proof. It readily follows that the support of $k_0 * a$ is contained in the ball $B(x_0, 2^4 r)$, and that the support of $k_i * a$ is disjoint with this ball as long as $i \ge 4$. Hence, $|k_0 * a(x)| \le |Ta(x)| + \sum_{i=1}^3 |k_i * a(x)|$. Since T is bounded on $L^{p_0}(R^n)$ we have $\|Ta\|_{L^{p_0}} \le c\|a\|_{L_0^p} \le cr^{n/p_0}$. Moreover, by Lemma 2, $\|k_i * a\|_{L^{p_0}} \le cr^{n/p_0}$, $i = 1, 2, \dots$, as well. Whence, $\|k_0 * a\|_{L^{p_0}} \le cr^{n/p_0}$. Since moments are not disturbed by convolutions, it follows that $k_0 * a$ is a (p_0, N_0) atom after normalization by a constant $\tilde{c}_0$, independent of r. ■

We remark that the assumption that T is bounded on $L^{p_0}(R^n)$ is trivially fulfilled when $p_0 = 2$. When $1 < p_0 \ne 2 < \infty$, we may use Theorem 4 below to verify that T is bounded on $L^{p_0}(R^n)$; this is possible since Lemma 2 is only used in the proof of Theorem 4 with $p_0 = 2$. Another way to arrive at the conclusion of Lemma 3 is to apply Theorem 4 directly to the multiplier $m * \phi_0$.

We consider now the multiplier result in the unweighted case.

Theorem 4. Let $m \in M(q, \ell)$, $1 \le q \le 2$ and $\ell > n/q$. Then, the associated mapping $Tf(x) = (f*k)(x)$, defined a-priori for $f \in \hat{\mathcal{D}}_0(R^n)$ extends to a bounded mapping from $L^p(R^n)$ into itself for $1 < p < \infty$. Also T maps $L^\infty(R^n)$ into $BMO(R^n)$.

Proof. We do the $L^p(R^n)$ case first. As is well-known, it suffices to consider the case $1 < p \le 2$. The case $p = 2$ is trivial since the $L^2(R^n)$ multipliers are the bounded functions and in this case the norm of the multiplier operator is $\|m\|_{L^\infty}$. Next consider the atomic decomposition of $f \in \hat{\mathcal{D}}_0(R^n)$ into $(2,0)$ atoms given in Theorem 1 of Chapter VIII, i.e., $f = \sum_j \lambda_j a_j$, where the a_j's are $(2,0)$ atoms with support contained in $B_j = B(x_j, r_j)$, $j \ge 0$, and $\|\sum_j \lambda_j a_j\|_{L^p} \le c\|f\|_{L^p}$; the convergence is taken here in the sense of the $L^p(R^n)$-norm. We want to show that

$$\left\| T\left(\sum_{j=1}^N \lambda_j a_j \right) \right\|_{L^p} \le c \left\| \sum_{j=1}^N \lambda_j a_j \right\|_{L^p} \le c\|f\|_{L^p}, \qquad (3)$$

with a constant c independent of N. From this estimate we conclude, by a limiting argument, that $\|Tf\|_{L^p} \le c\|f\|_{L^p}$ for $f \in \hat{\mathcal{D}}_0$.

Proof of (3). We make use of the estimates obtained in Lemmas 2 and 3 on each atom a_j, and split the kernel k according to the size of the balls B_j that contain the supports of the a_j's, to wit, $k = \sum_{i=0}^\infty k_{ij}$, say. Then we get

$$Ta_j = \sum_{i=0}^\infty k_{ij} * a_j = \sum_{i=0}^\infty (1/\tilde{c}_i) b_{ij}, \quad j = 1, \dots, N$$

where $b_{ij} = \tilde{c}_i(k_{ij} * a_j)$ is a $(2,0)$ atom with support contained in $B_{ij} = B(x_j, 2^{i+4} r_j)$; the hypothesis of Lemma 3 are fulfilled since T is bounded in $L^2(R^n)$. Whence, $T(\sum_{j=1}^N \lambda_j a_j) = \sum_{j=1}^N \sum_{i=0}^\infty (\lambda_j/\tilde{c}_i) b_{ij}$. Now, by Lemmas 2 and 3 in Chapter VIII it follows that

$$\left\| \sum_{j=1}^N (\lambda_j/\tilde{c}_i) b_{ij} \right\|_{L^p} \le \left\| \sum_{j=1}^N (\lambda_j/\tilde{c}_i) \chi_{B_{ij}} \right\|_{L^p}$$

$$\le c(2^{in}/\tilde{c}_i) \left\| \sum_{j=1}^N \lambda_j \chi_{B_j} \right\|_{L^p}.$$

By summation, then,

$$\left\| T\left(\sum_{j=1}^N \lambda_j a_j \right) \right\|_{L^p} \le c \left\| \sum_{j=1}^N \lambda_j \chi_{B_j} \right\|_{L^p} \sum_{i=0}^\infty (2^{in}/\tilde{c}_i).$$

But from Lemmas 2 and 3 we have that $\tilde{c}_i > c2^{i(n+\varepsilon)}$, where $c > 0$ is independent of i and $\varepsilon > 0$. Thus the sum over i converges, and the proof of (3) is complete.

Limiting argument. Let $f \in \hat{\mathcal{D}}_0(R^n)$; we want to show that $\|Tf\|_{L^p} \le c\|f\|_{L^p}$. Let $f = \sum_j \lambda_j a_j$ denote the atomic decomposition of f into $(\infty, 0)$ atoms with the property that if supp $a_j \subseteq B_j$, then $\sum_j \lambda_j \chi_{B_j}(x) \leq cf^*(x)$, and so that the finite sums $f_N = \sum_{j=1}^N \lambda_j a_j$ converge to f in the $L^p(R^n)$-norm. Since $f^* \in L^p(R^n)$, by the dominated convergence theorem it follows that the sequence $\left\{\sum_{j=1}^N \lambda_j \chi_{B_j}\right\}_{N=1}^\infty$ is Cauchy in $L^p(R^n)$, and then by (3) also the sequence $\{Tf_N\}_{N=1}^\infty$ is Cauchy in $L^p(R^n)$. Thus Tf_N converges, in the $L^p(R^n)$-norm, to some function g, say, in $L^p(R^n)$. On the other hand, since $f^* \in L^p(R^n)$, and since $|f_N| \leq cf^*$ and f_N converges to f in $L^p(R^n)$, also f_N converges to f in $L^2(R^n)$, again by the dominated convergence theorem. It then follows from the $L^2(R^n)$ estimate that Tf_N converges to Tf in $L^2(R^n)$. Thus Tf_N converges in measure, on compact sets, both to the functions Tf and g, and we conclude that $Tf = g$ a.e. Thus, $\lim_{N\to\infty} Tf_N = Tf$, in the sense of $L^p(R^n)$, and

$$\|Tf\|_{L^p} = \lim_{N\to\infty} \|Tf_N\|_{L^p} \leq c \lim_{N\to\infty} \left\|\sum_{j=1}^N \lambda_j \chi_{B_j}\right\|_{L^p}$$
$$\leq c\|f^*\|_{L^p} \leq c\|f\|_{L^p}.$$

This completes the proof of the $L^p(R^n)$ estimate, $1 < p < 2$. We observe that the above limiting argument works for any function $f \in L^p(R^n) \cap L^2(R^n)$ as well, where Tf is defined by $(Tf)\hat{}(\xi) = m(\xi)\hat{f}(\xi)$.

Next we consider the $L^\infty(R^n)$ case. The proof follows along similar lines, and we shall be brief. Let now $f \in L^\infty(R^n)$. It suffices to show that to each ball B there corresponds a constant c_B such that

$$I = \frac{1}{|B|} \int_B |Tf(x) - c_B|\, dx \leq c\|f\|_{L^\infty},$$

with c independent of f and B. By translation and dilation invariance we may assume that $B = B(0,1)$ is the unit ball centered at the origin. Let $2^j B = B(0, 2^j)$, and write $f = \sum_{j=0}^\infty f_j$, where $f_0 = f\chi_{2B}$ and $f_j = f(\chi_{2^{j+1}B} - \chi_{2^j B})$, $j = 1, 2, \ldots$ Then

$$I_0 = \int_B |f_0 * k(x)|\, dx \leq c\|f_0 * k\|_{L^2} \leq c\|f_0\|_{L^2} \leq c\|f\|_{L^\infty}.$$

Similarly, if $c_{B,j} = \int_{R^n} f_j(y) k(-y)\, dy$, $j \geq 1$, we have that

$$I_j = \int_B |f_j * k(x) - c_{B,j}|\, dx$$
$$\leq c \left(\int_B \left| \int_{R^n} f_j(y)(k(x-y) - k(-y))\, dy \right|^{p'} dx \right)^{1/p'}$$
$$\leq c \left(\int_B \left(\int_{R^n} |f_j(y)|^p \right)^{p'/p} \left(\int_{\{|y|\sim 2^j\}} |k(x-y) - k(-y)|^{p'}\, dy \right)^{p'} \right)^{1/p'}.$$

Note that the first integral above is bounded by $c2^{jn/p}\|f\|_{L^\infty}$. Also, from Lemma 2, we see that the second integral above does not exceed $c^{p'}2^{-j(1+n/p)}$. Whence, $I_j \le c2^{-j}\|f\|_{L^\infty}$, and, as assserted,

$$I \le \sum_{j=0}^{\infty} I_j \le c\|f\|_{L^\infty}. \quad \blacksquare$$

We can now state the multiplier results for the weighted Lebesgue spaces.

Theorem 5. Let ν be a doubling measure with respect to the Lebesgue measure on R^n, $d\nu(x) = w(x)dx$, and suppose $w \in A_s \cap RH_r$, $r,s > 1$. Further, let $1 \le q \le 2$, $s \le p \le \infty$, and suppose that $m \in M(q,\ell)$, where $\ell > n/q$ and $\ell \ge n\min(s/p, 1/p' + 1/(rp))$. Then the multiplier operator associated with m is a bounded mapping from $L_w^p(R^n)$ into itself.

By a duality argument and general properties of the weights, the proof of Theorem 5 is reduced to

Theorem 5, reduced version. Let ν be a doubling weighted measure with respect to the Lebesgue measure on R^n, $d\nu(x) = w(x)dx$, and assume that the weight $w \in A_p \cap RH_r$. Let $1 \le q \le 2$, and suppose $m \in M(q,\ell)$, where ℓ is such that $\ell > n\max(1/q, 1/p' + 1/(rp))$. Then, the multiplier operator associated to m is a bounded mapping from $L_w^p(R^n)$ into itself.

Assuming the reduced version of Theorem 5 for the moment, we will see how the general version follows. First, we may assume that the inequality $\ell > n\min(s/p, 1/p' + 1/(rp))$ in Theorem 5 is strict since $w \in A_s \cap RH_r$ implies that $w \in A_{s-\varepsilon} \cap RH_{r+\varepsilon}$ for some $\varepsilon > 0$. Next, we observe that under the hypothesis of Theorem 5, the reduced version applies to the space $L_w^p(R^n)$ and its dual $L_{w^{-1/(p-1)}}^{p'}(R^n)$. To see this we only need to check that $w \in A_s$ implies that $w^{-1/(p-1)} \in A_{p'} \cap RH_{r_1}$, with $r_1 = (p-1)/(s-1)$ and $1/p + 1/r_1p' = p/s$. In this case, by a duality argument, it follows that a multiplier that is bounded on $L_{w^{-1/(p-1)}}^{p'}(R^n)$, is also bounded on $L_w^p(R^n)$.

We may now proceed with

Proof of Theorem 5, reduced version. We show that $\|Tf\|_{L_w^p} \le c\|f\|_{L_w^p}$ for $f \in \hat{\mathcal{D}}_0$; the operator T then extends to a bounded mapping on $L_w^p(R^n)$.

The proof is very similar to that of Theorem 4, and we indicate the differences with that proof here. Let $f \in \hat{\mathcal{D}}_0$. We then decompose f into a sum of $(\infty,0)$ atoms such that the corresponding sum of the characteristic functions $\sum_j \lambda_j \chi_{B_j}$ is bounded in the $L_w^p(R^n)$-norm by $\|f\|_{L_w^p}$ and $\lim_{N\to\infty}\|\sum_{j=1}^N \lambda_j a_j - f\|_{L_w^p} = 0$. Next we choose $p_0 > pr/(r-1)$ and $\ell > n(1 - 1/p_o)$; the last choice is possible since $\ell > n(1 - (r-1)/pr)$. Since the $(\infty, 0)$ atoms a_j are $(p_0, 0)$ atoms as well, and since by Theorem 4 the operator T is bounded on $L^{p_0}(R^n)$, we may apply Lemmas 2 and 3 on each atom a_j. In this way $T(\sum_{j=1}^{N} \lambda_j a_j)$ can be written as a double sum $\sum_{j=1}^{N} \sum_{i=0}^{\infty} (\lambda_j / \tilde{c}_i) b_{ij}$, where the b_{ij}'s are $(p_0, 0)$ atoms with support contained in $2^{i+4} B_j$. By Lemma 4 in Chapter VIII we get

$$\left\| \sum_{j=1}^{N} (\lambda_j / \tilde{c}_i) b_{ij} \right\|_{L_w^p} \leq c \left\| \sum_{j=1}^{N} (\lambda_j / \tilde{c}_i) \chi_{2^{i+4} B_j} \right\|_{L_w^p}.$$

It is at this point that we invoke the fact that $w \in RH_r$ and $p_0 > pr/(r-1)$. Now, by Lemma 3 in Chapter VIII, since $w \in A_p$, this last norm is dominated by

$$c(2^{in}/\tilde{c}_i) \left\| \sum_{j=1}^{N} \lambda_j \chi_{B_j} \right\|_{L_w^p}.$$

As pointed out in Lemma 2, we have $\tilde{c}_i \geq c 2^{i(n+\varepsilon)}$ for some $\varepsilon > 0$ when $\ell > n \max(1/q, 1 - 1/p)$. Summing over i we obtain

$$\left\| T\left(\sum_{j=1}^{N} \lambda_j a_j \right) \right\|_{L_w^p} \leq c \left\| \sum_{j=1}^{N} \lambda_j \chi_{B_j} \right\|_{L_w^p} \leq c \|f\|_{L_w^p}.$$

As for the limiting argument needed to obtain the $L_w^p(R^n)$-norm estimate for Tf, it is the same as that in the proof of Theorem 4 and is therefore omitted. This completes the proof of Theorem 5 in the general and reduced versions. ∎

We now state the multiplier result for the weighted Hardy spaces. First we recall that if $w \in A_s$ for some $s \geq 1$ and $p \geq s$, then $H_w^p(R^n)$ is identical to $L_w^p(R^n)$, and in this case Theorem 5 applies.

When $0 < p < s$ we have the following result.

Theorem 6. Let ν be a doubling weighted measure with respect to the Lebesgue measure on R^n, $d\nu(x) = w(x)dx$, and suppose that $\nu \in D_{n\theta}$ with $1 \leq \theta \leq s$. Further, let $1 \leq q \leq 2$ and $0 < p < s$ and assume $m \in M(q, \ell)$, where

$$\ell \begin{cases} \geq \dfrac{n(\theta-1)(s-p)}{p(s-1)} + \max\left(\dfrac{1}{q}, \dfrac{1}{p'} + \dfrac{1}{rp}\right) & \text{if } 1 < p < s \\[2mm] > n\left(\dfrac{\theta}{p} - \dfrac{1}{q'}\right) \text{ and } \geq n\dfrac{1}{p}\left(\theta - \dfrac{r-1}{r}\right) & \text{if } 0 < p \leq 1. \end{cases}$$

Then, the multiplier operator T associated to m is a bounded mapping of $H_w^p(R^n)$ into itself.

A word about this result. We recall that $w \in A_s$ implies that $\nu \in D_{ns}$, and that $w \in A_\infty$ is equivalent to $w \in A_s$ for some $s > 1$, and $w \in RH_r$ for some $r > 1$. For instance, if we know that $w \in A_s$, then we must have $\ell \geq ns/p$ in order to apply Theorem 5. On the other hand, if we only know that $\nu \in D_{n\theta}$ and $w \in A_\infty$, we need to have $\ell \geq n(\theta-1)/p + n$ in case $1 < p < \infty$, and $\ell \geq n\theta/p$, with strict inequality when $p = 1$, in case $0 < p \leq 1$.

Proof. We may assume that all three inequalities satisfied by ℓ are strict, since we can always increase r and decrease s a little, and still have $w \in A_s \cap RH_r$. We have to show an $H_w^p(R^n)$-norm estimate for Tf, where $f \in \hat{\mathcal{D}}_0$. As before, we decompose f into a sum of atoms and estimate T applied to finite sums of atoms by applying Lemmas 2 and 3 on each atom to get, in each instance, a double sum which is handled as in Theorem 5. The main difference from the proof of that theorem is that the atoms need, in general, satisfy moment conditions of order higher than 0. This fact is used in three places in the proof. First, to get the finite sums of atoms to converge to f in the $H_w^p(R^n)$-norm. Second, when we estimate the $H_w^p(R^n)$-norm of $\sum_{j=1}^N (\lambda_j/\tilde{c}_i)b_{ij}$ in terms of the $L_w^p(R^n)$-norm of $\sum_{j=1}^N (\lambda_j/\tilde{c}_i)\chi_{2^{i+4}B_j}$. Both of these instances use Theorem 1 in Chapter VIII, where, in general, higher order moments are required. The third place is in Lemma 2, to get the constants $\tilde{c}_i$ to increase rapidly enough: When estimating

$$\left\| \sum_{j=1}^N (\lambda_j/\tilde{c}_i)\chi_{2^{i+4}B_j} \right\|_{L_w^p} \quad \text{by} \quad \left\| \sum_{j=1}^N \lambda_j \chi_{B_j} \right\|_{L^p w}$$

using Lemma 3 in Chapter VIII, we get a factor $c(2^{i\delta}/\tilde{c}_i)$, and in order to get a convergent sum over i we need $\tilde{c}_i \geq c2^{i(\delta+\varepsilon)}$ for some $\varepsilon > 0$, and $c > 0$. The order of the moment condition needed is directly related to this number δ, also when applying Lemma 3 in Chapter VIII. More precisely, we need to decompose f into (∞, N_0) atoms, with $N_0 > \delta - n - 1$, and apply Lemmas 2 and 3 on (p_0, N_0) atoms with $p_0 > pr/(r-1)$. We recall that we have $\delta = n\theta/p$ when $0 < p \leq 1$, and $\delta = n + n(\theta - 1)(s - p)/p(s - 1)$ when $1 < p < s$. We can choose p_0 such that $\ell > \delta - n \min(1/q', 1/p_0)$. With the notation of Lemma 2 it follows that $\tilde{\ell} > \delta - n$, and we conclude from this lemma that $\tilde{c}_i \geq c2^{i(\delta+\varepsilon)}$ for some $\varepsilon > 0$, and a constant $c > 0$. This is precisely what we need.

The limiting argument is quite similar to that given in the proof of Theorem 4. This time the finite sum of atoms f_N is a Cauchy sequence in $H_w^p(R^n)$. By the a-priori estimate of T acting on finite sums of atoms, Tf_N is also a Cauchy sequence converging, in the $H_w^p(R^n)$-norm, to some distribution g. As before we can show that Tf_N also converges to Tf in the $L^2(R^n)$-norm. This means that Tf_N converges to both g and Tf in the distribution sense. Thus, $g = Tf$, and $\|Tf\|_{H_w^p} \leq c\|f\|_{H_w^p}$, and the proof is complete. ∎

Alternate Proof of Theorem 6, $0 < p \leq 1$. Instead of using the full strength of Theorem 4, second half, and Lemma 3 in Chapter VIII, we only need to invoke the following estimate on a single atom, namely,

$$\|a\|_{H_w^p} \leq c\nu(B)^{1/p},$$

for any (p_0, N_0) atom a with support contained in the ball B, provided that $p_0 > pr/(r-1)$, $N_0 > (n\theta/p) - n - 1$, and $\nu \in D_{n\theta}$ and $w \in RH_r$. Under these

Let, then, $f \in \hat{\mathcal{D}}_0(R^n)$. As before we decompose f into a sum $\sum_j \lambda_j a_j$ where the a_j's are (∞, N) atoms with support contained in the balls B_j, say, and such that the partial sums $\sum_{j=1}^N \lambda_j a_j$ converge to f in the $H_w^p(R^n)$-norm. As noted in Chapter VIII, this can be done in a such a way that $\sum_j \lambda_j^p \nu(B_j) \leq c\|f\|_{H_w^p}^p$. As before we also get that $T(\sum_{j=1}^N \lambda_j a_j) = \sum_{i=0}^\infty \sum_{j=1}^N (\lambda_j/\tilde{c}_i)b_{ij}$, where the b_{ij}'s are (p_0, N_0) atoms with support contained in $2^{i+4}B_j$. We then have

$$\|b_{ij}\|_{H_w^p}^p \leq c\nu(2^{i+4}B_j) \leq c2^{in\theta}\nu(B_j),$$

provided $p_0 > pr/(r-1)$ and $N_0 > (n\theta/p) - n - 1$. By the triangle inequality it follows that

$$\left\|\sum_{j=1}^N \sum_{i=0}^\infty (\lambda_j/\tilde{c}_i)b_{ij}\right\|_{H_w^p}^p \leq \sum_{j=1}^N \sum_{i=0}^\infty (\lambda_j/\tilde{c}_i)^p \|b_{ij}\|_{H_w^p}^p$$
$$\leq \sum_{i=0}^\infty (2^{in\theta}/\tilde{c}_i^p) \sum_{j=1}^N \lambda_j^p \nu(B_j) \leq c \sum_{i=0}^\infty (2^{in\theta}/\tilde{c}_i^p)\|f\|_{H_w^p}^p.$$

As before we can pick p_0 so that $\ell > (n\theta/p) - n \min(1/q', 1/p_0)$ and obtain, by Lemmas 2 and 3, $\tilde{c}_i \geq c2^{i((n\theta/p)+\varepsilon)}$ for some $\varepsilon > 0$, $c > 0$. We conclude then, that $\|T(\sum_{j=1}^N \lambda_j a_j)\|_{H_w^p} \leq c\|f\|_{H_w^p}$. A limiting argument gives now $\|Tf\|_{H_w^p} \leq c\|f\|_{H_w^p}$, and this concludes the proof. ∎

An interesting result in the unweighted case is the following corollary to Theorem 6.

Corollary. Let $1 \leq q \leq 2$, $0 < p \leq 1$, and suppose that $m \in M(q, \ell)$, where $\ell > n(1/p - 1/q')$. Then the multiplier operator associated to m is a bounded mapping from $H^p(R^n)$ into itself.

We pass now to discuss a different approach to deal with multipliers, namely, that of pointwise comparison of sharp maximal functions. For reasons of clarity and emphasis we depart from the notation introduced in Chapter III and, given a weighted measure ν with respect to the Lebesgue measure on R^n, $d\nu(x) = w(x)dx$, we put

$$M_{r,\nu}^{\sharp,p} f(x) = \sup_{B \supset \{x\}} \inf_{p \in \mathcal{P}} \left(\frac{1}{\nu(B)} \int_B |f(y) - p(y)|^r w(y)\, dy\right)^{1/r}.$$

We prove our results only for singular integral operators. This is done for simplicity and the reader should have no difficulty in extending them for general multipliers.

Proposition 7. Let ν be a doubling weighted measure with respect to the Lebesgue measure on R^n, $d\nu(x) = w(x)dx$, and suppose $\nu \in D_{n\theta}$ and $w \in RH_r$, where $r, \theta > 1$. Further, let the kernel $k \in \tilde{M}(\tilde{q}, \tilde{\ell})$, $1 \le \tilde{q} \le \infty$, and assume the associated operator T satisfies

$$\int_B |Tf(x)|^{q_0}\, dx \le c|B|\,, \quad 0 < q_0 < \infty\,, \quad \text{or} \quad |Tf(x)| \le c \quad \text{if} \quad q_0 = \infty\,,$$

for all balls B and all bounded functions f supported in B with $\|f\|_\infty \le 1$. Finally, let $0 \le d_1 < \infty$, $\min(1, d_1) \le d_2 < \infty$, and also $d_1 \le \min(\tilde{q}/r', q_0/r')$.
Then, we have

$$M_{d_1,\nu}^{\sharp,\mathcal{P}}(Tf)(x) \le c M_{d_2,\nu}(f^*)(x)\,, \quad f \in \hat{\mathcal{D}}_0\,,$$

provided $\tilde{\ell} > n\theta/\min(\tilde{q}/r', d_2, 1) - n/\tilde{q}'$ and $\mathcal{P}$ is the class of polynomials on degree less than or equal to N_0, where N_0 is greater or equal to the integer part of $\tilde{\ell}$.

A remark before we proceed, namely, the assumption on Tf holds if T is bounded on $L^{q_0}(R^n)$ or more generally if T is of weak-type (q, q) for some $q > q_0$.

Proof. By using Hölder's inequality in the definition of the various maximal functions involved we may assume that $d_1 > 0$ and $d_2 \le \min(\tilde{q}/r', 1)$. Thus, we have two cases to consider, namely, $0 < d_1 \le d_2 \le 1$, and $1 = d_2 < d_1 < \infty$. Set $d_3 = \max(d_1, d_2)$. From the hypothesis we may assume that $\tilde{q} < \infty$ and that $\tilde{\ell}$ is not an integer. This will facilitate the writing of the proof although still the simplest case is $\tilde{q} = \infty$. We may also assume that $q_0 < \infty$.

By a translation and dilation argument it suffices to estimate

$$\frac{1}{\nu(B)} \int_B |Tf(x) - p(x)|^{d_1} w(x)\, dx$$

where $B = B(0, 1)$ denotes the unit ball in R^n and $p \in \mathcal{P}$ is a suitable polynomial of degree less than or equal to $N_0 = $ integer part of $\tilde{\ell}$. According to the atomic decomposition we write $f = \sum_j \lambda_j a_j$ as the sum of (∞, N_0) atoms a_j with support contained in balls B_j and so that the corresponding sum of characteristic functions of these balls satisfies

$$\sum_j \lambda_j^d \chi_{B_j}(x) \le c_d f^*(x)^d\,, \quad \text{all } 0 < d \le 1\,.$$

When $d_1 = d_3 = d$ we proceed to estimate

$$\left(\frac{1}{\nu(B)} \int_B |Ta_j(x) - p_j(x)|^d w(x)\, dx \right)^{1/\max(d,1)}$$

$$\le c \sup_{i \ge 0} 2^{-i\varepsilon} \frac{1}{\nu(2^i B)} \int_{2^i B} \chi_{B_j}(x) w(x)\, dx\,,$$

for some $\varepsilon > 0$, and we then get the desired estimate by summation; here p_j is a suitable polynomial of degree less than or equal to N_0.

More generally, when $d_1 \leq d_3$, the idea is to split $Ta_j(x)$ into two functions, $Ta_j(x) = \Phi_j(x) + \Psi_j(x)$, say, and to estimate

$$\left(\frac{1}{\nu(B)} \int_B |\Phi_j(x)|^{d_1} w(x) \, dx \right)^{1/\max(d_1,1)}$$

and

$$\left(\frac{1}{\nu(B)} \int_B |\Psi_j(x) - p_j(x)|^{d_3} w(x) \, dx \right)^{1/\max(d_3,1)},$$

by the right-hand side above. In this case $Tf(x) = \sum_j \lambda_j \Phi_j(x) + \sum_j \lambda_j \Psi_j(x)$, and with $p(x) = \sum_j \lambda_j p_j(x)$ we get

$$\left(\frac{1}{\nu(B)} \int_B |Tf(x) - p(x)|^{d_1} w(x) \, dx \right)^{1/\max(d_1,1)}$$

$$\leq \sum_j \lambda_j^{\min(d_1,1)} \left(\frac{1}{\nu(B)} \int_B |\Phi_j(x)|^{d_1} w(x) \, dx \right)^{1/\max(d_1,1)}$$

$$+ \left(\sum_j \lambda_j^{\min(d_3,1)} A_j(B) \right)^{\max(d_3,1)/\min(d_1,1)}, \tag{4}$$

Here we used Hölder's inequality, the triangle inequality and the notation

$$A_j(B) = \left(\frac{1}{\nu(B)} \int_B |\Psi_j(x) - p_j(x)|^{d_3} w(x) \, dx \right)^{1/\max(d_3,1)}.$$

Assume for the moment that the integrals on the right-hand side above can be estimated as indicated. Then the first sum on the right-hand side of (4) is bounded by

$$\sum_j \lambda_j^{\min(d_1,1)} \sup_{i \geq 0} 2^{-i\varepsilon} \frac{1}{\nu(2^i B)} \int_{2^i B} \chi_{B_j}(x) w(x) \, dx$$

$$\leq \sum_{i \geq 0} 2^{i\varepsilon} \frac{1}{\nu(2^i B)} \int_{2^i B} \sum_j \lambda_j^{\min(d_1,1)} \chi_{B_j}(x) w(x) \, dx$$

$$\leq c \sup_{i \geq 0} \frac{1}{\nu(2^i B)} \int_{2^i B} f^*(x)^{\min(d_1,1)} w(x) \, dx.$$

In a similar fashion, the second sum on the right-hand side of (4) is bounded by

$$c \sup_{i \geq 0} \frac{1}{\nu(2^i B)} \int_{2^i B} f^*(x)^{\min (d_3,1)} w(x)\, dx\,.$$

Combining these estimates it then follows that $M_{d_1,\nu}^{\sharp,P}(Tf)(x) \leq cM_{d_3,\nu}(f^*)(x)$.

The main part of the proof still remains to be carried out, namely, we must define the function $\Phi_j(x)$ and $\Psi_j(x)$ and show they satisfy the desired properties.

Let r_j denote the radius of B_j. We then let

$$\Phi_j(x) = \chi_{2B_j}(x)Ta_j(x)\,, \quad \Psi_j(x) = (1 - \chi_{2B_j}(x))Ta_j(x) \quad \text{when } r_j \leq 1$$

and

$$\Phi_j(x) = T(\chi_{2B}a_j)(x)\,, \quad \Psi_j(x) = T((1 - \chi_{2B})a_j)(x) \quad \text{when } r_j \geq 1\,.$$

We want to show the estimates

$$\left(\frac{1}{\nu(B)} \int_B |\Phi_j(x)|^{d_1} w(x)\, dx\right)^{1/\max (d_1,1)}$$

$$\leq c \sup_{i \geq 0} 2^{-i\varepsilon} \frac{1}{\nu(2^i B)} \int_{2^i B} \chi_{B_j}(x)w(x)\, dx \qquad (5)$$

and

$$\left(\frac{1}{\nu(B)} \int_B |\Psi_j(x) - p_j(x)|^{d_3} w(x)\, dx\right)^{1/\max (d_3,1)}$$

$$\leq c \sup_{i \geq 0} 2^{-i\varepsilon} \frac{1}{\nu(2^i B)} \int_{2^i B} \chi_{B_j}(x)w(x)\, dx\,, \qquad (6)$$

for some $\varepsilon > 0$ and for some polynomial p_j of degree less than or equal to N_0 depending on B and a_j.

Let j be fixed, and let i_0 be the smallest integer i such that $2^i B \cap B_j \neq \emptyset$. First we estimate the right-hand sides of (5) and (6). When $r_j \geq 2^{i_0}$, since there is a ball $\tilde{B}_j$ in $B_j \cap 2^{i_0+1}B$ with $2^{i_0+1}B \subseteq c_0\tilde{B}_j$ for some dimensional constant c_0, these quantities are at least greater than or equal to

$$c2^{-i_0}\nu(B_j \cap 2^{i_0+1}B)/\nu(2^{i_0+1}) \geq c2^{-i_0\varepsilon}\,.$$

Also, when $r_j < 2^{i_0}$, B_j is contained in $2^{i_0+1}B$, so that the right-hand sides of (5) and (6) are larger than

$$c2^{-i_0\varepsilon}\nu(B_j)/\nu(2^{i_0+2}B) \geq c2^{-i_0(\varepsilon+n\theta)}r_j^{n\theta}\,.$$

Next we estimate the left-hand sides of (5) and (6); we do (5) first. We consider several cases. When $i_0 > 1$ it is easy to check that $\Phi_j(x)$ vanishes identically on B. When $i_0 = 1$ and $r_j \leq 1$, we use the reverse Hölder's condition on the ball $2B_j$ and the fact that $\int_{2B_j} |Ta_j(x)|^{q_0} dx \leq c|2B_j|$ to get that $\int_B |\Phi_j(x)|^{d_1} w(x) dx \leq c\nu(2B_j)$ and thus the left-hand side of (5) is less than or equal to

$$c(\nu(2B_j)/\nu(B))^{1/\max(d_1,1)} \leq c\nu(B_j)/\nu(8B),$$

which in turn is dominated by the right-hand side of (5). Next, the case $i_0 = 1$ and $r_j > 1$ is very similar. We use the reverse Hölder's condition on the ball B and the fact that $\int_{2B} |T(\chi_{2B} a_j)(x)|^{q_0} dx \leq c|B|$, to get that

$$\int_B |\Phi_j(x)|^{d_1} w(x) dx \leq c\nu(B).$$

Thus, the left-hand side of (5) is bounded above by a constant, while the right-hand side of (5) is bounded below by a positive constant.

In order to estimate (6) we consider the Taylor expansion of order $N_0 - 1$ of $k(x - z)$. When $r_j \leq 1$ we expand $k(x - z)$ as a function of z around x_j, the center of B_j. We then have $k(x - z) = p_x(z) + R_x(z)$, where $p_x(z)$ is a polynomial in z of degree less than or equal to N_0, and

$$|R_x(z)| \leq c \sum_{|\beta|=N_0} \int_0^1 |D^\beta k((x - x_j) - s(z - x_j)) - D^\beta k(x - x_j)| \, |x - x_j|^{N_0} ds.$$

The assumption $k \in \tilde{M}(\tilde{q}, \tilde{\ell})$ implies that

$$\left(\int_{\{|x-x_j| \sim 2^i\}} |R_x(z)|^{\tilde{q}} dx \right)^{1/\tilde{q}} \leq c2^{-in/\tilde{q}'}(2^{-i}r_j)^{\tilde{\ell}}$$

whenever $z \in B_j$ and $2^i > r_j$. So, using the moment condition on the atom a we get that $k * a_j(x) = \int_{R^n} R_x(z) a_j(z) dz$, and with $p_j(x) = 0$, by Hölder's inequality and the reverse Hólder's condition, we get

$$\frac{1}{\nu(B)} \int_B |\Psi_j(x) - p_j(x)|^{d_3} w(x) dx \leq c \left(\int_{B \setminus 2B_j} |k * a_j(x)|^{\tilde{q}} dx \right)^{d_3/\tilde{q}}.$$

Since $\operatorname{dist}(x_j, B \setminus 2B_j) \geq c2^{i_0}$, by Minkowki's inequality and the estimates on $R_x(z)$ above we have

$$\left(\int_{B \setminus 2B_j} |k * a_j(x)|^{\tilde{q}} dx \right)^{1/\tilde{q}} \leq c2^{-i_0 n/\tilde{q}'}(2^{-i}r_j)^{\tilde{\ell}}|B_j|.$$

Thus, observing that $d_3/\max(d_3,1) = d_2$, we conclude that the left-hand side of (6) is bounded by $c2^{-i_0 d_2((n/\tilde{q}')+\tilde{\ell})} r_j^{d_2(\tilde{\ell}+n)}$, which can in turn be dominated by the right-hand side of (6) if $d_2(\tilde{\ell} + (n/\tilde{q}')) \geq n\theta + \varepsilon$. From our assumptions, this inequality holds provided $\varepsilon > 0$ is small enough.

When $r_j > 1$ we expand $k(x-z)$ as a function of x about the origin and write $k(x-z) = p_z(x) + R_z(x)$, where $p_z(x)$ is a polynomial in x of degree less than or equal to N_0, and

$$\left(\int_{\{|z|\sim 1\}} |R_z(x)|^{\tilde{q}} \, dz \right)^{1/\tilde{q}} \leq c2^{-i((n/q')+\tilde{\ell})}$$

when $i > 0$ and $x \in B$. We define the polynomials $p_j(x)$ then by

$$p_j(x) = \int_{R^n} p_z(x)(1 - \chi_{2B}(z)) a_j(z) \, dz \,,$$

and get $\Psi_j(x) = \int_{R^n} R_z(x)(1 - \chi_{2B}(z)) a_j(z) \, dz$. By a simple computation using Hölder's inequality on the sets $\{|z| \sim 2^i\}$ and summation we get

$$|\Psi_j(x) - p_j(x)| \leq c \int_{B_j \setminus 2B} |R_z(x)| \, dz \leq c2^{i_0 \tilde{\ell}}, \quad x \in B \,.$$

We thus conclude that the left-hand side of (6) is dominated by $c2^{-i_0 d_2 \tilde{\ell}}$, and that (6) holds when $r_j \geq 2^{i_0}$ provided we choose ε so that $0 < \varepsilon < \tilde{\ell}$. In case $1 < r_j < 2^{i_0}$, by Hölder's inrequality it follows that

$$|\Psi_j(x) - p_j(x)| \leq \left(\int_{B_j \setminus 2B} |R_z(x)|^{q'} \, dz \right)^{1/q'} |B_j \setminus B|^{1/q'}$$

$$\leq c2^{i_0(\tilde{\ell}+(n/\tilde{q}'))} r_j^{n/\tilde{q}'}, \quad x \in B \,.$$

Thus, the left-hand side of (6) is bounded by the above expression raised to the power d_2, and we conclude that (6) holds in this case provided that ε is chosen so that $d_2(\tilde{\ell} + (n/\tilde{q}')) > n\theta + \varepsilon$. The proof is thus complete. ∎

Proposition 7 leads to weighted norm inequalities for the singular integral operator T. Let ν be a doubling weighted measure with respect to the Lebesgue measure on R^n, $d\nu(x) = w(x)dx$, and assume that $\nu \in D_{n\theta}$ and $w \in A_s \cap RH_r$. We will apply Proposition 7 to the weighted measure ν_b defined by $d\nu_b(x) = w(x)^b dx$, where $0 \leq b \leq 1$. Observe that $\nu_b \in D_{n\theta_b}$ and $w^b \in RH_{r_b}$, where $\theta_b = b(\theta-1)+1$ and $r_b = r/b$. Also $w \in A_{s_b}(\nu_b)$, where $1/s_b = 1/s + b/s'$. Now, from Theorem 2 in Chapter III it follows that there is a polynomial p of degree less than or equal to N_0, depending on $f \in \hat{\mathcal{D}}_0$, such that

$$\|Tf - p\|_{L^p_w} \le c\|M^{\sharp,\mathcal{P}}_{0,\nu,s_1}(Tf)\|_{L^p_w}, \quad \text{for } s_1 > 0 \text{ small enough}.$$

As we have seen in Chapter III, there is $s_2 > 0$ such that

$$M^{\sharp,\mathcal{P}}_{0,\nu,s}g(x) \le cM^{\sharp,\mathcal{P}}_{0,\nu_b,s_2}g(x) \le cs_2^{1/d_1} M^{\sharp,\mathcal{P}}_{d_1,\nu_b}g(x);$$

in the first two inequalities above we used that $w \in A_\infty$. This observation together with Proposition 7, give the pointwise estimate

$$M^{\sharp,\mathcal{P}}_{0,\nu,s}(Tf)(x) \le cM_{d_2,\nu_b}(f^*)(x), \quad f \in \hat{\mathcal{D}}_0,$$

provided k is as in Proposition 7 and $\tilde{\ell} > n\theta_b/\min{(\tilde{q}/r_b, d_2, 1)} - n/\tilde{q}'$. Moreover, since the maximal operator M_{d_2,ν_b} is bounded on $L^p_w(R^n)$, when $p/d_2 = s_b$, we get

$$\|Tf - p\|_{L^p_w} \le c\|f^*\|_{L^p_w}, \quad f \in \hat{\mathcal{D}}_0,$$

provided $\tilde{\ell} > n\theta_b/\min{(\tilde{q}/r_b, p/s_b, 1)} - n/\tilde{q}'$.

It is not difficult to check that the polynomial p above is zero when $f \in \hat{\mathcal{D}}_0$. Indeed, by direct estimates we get

$$\int_{\{|x|\sim R\}} |Tf(x)|^{p_0}\,dx = O(R^n), \quad \text{as} \quad R \to \infty,$$

and by Hölder's inequality it follows that

$$\int_{\{|x|\sim R\}} |Tf(x) - p(x)|^{p_0}\,dx = O(R^n), \quad \text{as} \quad R \to \infty,$$

when $\|Tf - p\|_{L^p_w} < \infty$, if p_0 is chosen small enough. Whence

$$\int_{\{|x|\sim R\}} |p(x)|^{p_0}\,dx = o(R^n), \text{as} \quad R \to \infty,$$

and consequently, $p(x) = 0$.

We have thus shown the estimate

$$\|Tf\|_{L^p_w} \le c\|f\|_{H^p_w}, \quad f \in \hat{\mathcal{D}}_0,$$

provided $\tilde{\ell} > n\theta_b/\min{(\tilde{q}/r_b, s_b, 1)} - n/\tilde{q}'$. Next we minimize this last expression by choosing appropriately b, $0 \le b \le 1$; we only consider some cases in the theorem below. Note that we may replace the $L^p_w(R^n)$-norm on the left-hand side above by the $H^p_w(R^n)$-norm of Tf. A way to see this is to express the $H^p_w(R^n)$-norm of Tf by mean of its $L^p_w(R^n)$-norm and that of its Riesz transforms of high enough order, assume that we know the fact that such transforms are bounded on $L^p_w(R^n)$, and then note that they commute with T.

In fact, choosing a suitable b this method gives

Theorem 8. Let ν be a doubling weighted measure with respect to the Lebesgue measure on R^n, $d\nu(x) = w(x)dx$, and assume that $\nu \in D_{n\theta}$ where $\theta \geq 1$ and $w \in A_s \cap RH_r$ where $1 < r, s < \infty$. Further, let $k \in \tilde{M}(\tilde{q}, \tilde{\ell})$, $1 \leq \tilde{q} \leq \infty$, denote the kernel of a singular integral operator T so that there is q_0, $0 < q_0 \leq \infty$, with the property that

$$\int_B |Tf(x)|^{q_0}\, dx \leq c|B|, \quad 0 < q_0 < \infty, \quad \text{or} \quad |Tf(x)| \leq c \quad \text{if} \quad q_0 = \infty,$$

for all balls B and all functions f supported in B such that $|f(x)| \leq 1$. Then,

$$\|Tf\|_{H^p_w} \leq c\|f\|_{H^p_w} \quad \text{for} \quad f \in \hat{\mathcal{D}}_0,$$

provided that
    (i) $\tilde{\ell} > n(\theta/p - 1/\tilde{q}')$ when $0 < p \leq 1$ and $\tilde{q}/r' \geq p$,
or
    (ii) $\tilde{\ell} > n\left(\frac{1}{s}\frac{(s-p)}{(s-1)}(\theta - 1) + 1/\tilde{q}'\right)$, when $s > p > 1$ and $\frac{1}{p}\frac{(s-p)}{(s-1)} \leq r/\tilde{q}'$.
    Furthermore,
$$\|Tf\|_{L^p_w} \leq c\|f\|_{L^p_w}, \quad f \in \hat{\mathcal{D}}_0,$$

(iii) provided $\tilde{\ell} > n/\tilde{q}$ when $p \geq s > 1$.

Proof. We need only pick b appropriately. For the cases (i), (ii) and (iii), we choose $b = 1$, $(1/p)(s-p)/(s-1)$, and 0, respectively. Also observe that $H^p_w(R^n)$ coincides with $L^p_w(R^n)$ when $p \geq s > 1$. ∎

For $L^p_w(R^n)$-norm inequalities with $w \in A_s$, $1 < s \leq p$, it is in some cases preferable to use the following estimate involving unweighted maximal functions; the estimate does not involve the atomic decomposition.

Proposition 9. Let $k \in \tilde{M}(\tilde{q}, \tilde{\ell})$, $1 < \tilde{q} \leq \infty$, $\tilde{\ell} > 0$, and suppose that the associated singular integral operator T preserves $L^{q_0}(R^n)$, $1 \leq q_0 < \infty$. If $0 \leq d_1 \leq q_0$ and $d_2 \geq \max(q_0, \tilde{q}')$, we have

$$M^{\sharp,\mathcal{P}}_{d_1}(Tf)(x) \leq cM_{d_2}f(x), \quad f \in \hat{\mathcal{D}}_0,$$

where $\mathcal{P} = \text{constants}$.

Sketch of the Proof. We may assume that $d_1 > 0$. We want to estimate

$$\frac{1}{|B|}\int_B |Tf(x) - c|^{d_1}\, dx$$

where $B = B(0,1)$ is the unit ball. Write $f = \sum_{i=0}^{\infty} f_i$, where supp $f_0 \subseteq 2B$ and supp $f_i \subseteq 2^{i+2}B \setminus 2^i B$ for $i = 1, 2, \ldots$ For Tf_0 we use the assumption that T

preserves $L^{q_0}(R^n)$, and for Tf_i we observe that with $c_i = -\int_{R^n} k(-z)f_i(z)\,dz$, we have

$$Tf_i(x) - c_i = \int_{R^n} (k(y-z) - k(-z))f_i(z)\,dz, \quad i = 1, 2, \ldots$$

Using Hölder's inequality and the condition $\tilde{M}(\tilde{q}, \tilde{\ell})$, it follows there exists $0 < \varepsilon < \min(\tilde{\ell}, 1)$, such that

$$|Tf_i(y) - c_i| \le c2^{-i\varepsilon} \left(2^{-in} \int_{2^{i+1}B} |f(z)|^{\tilde{q}'}\,dz \right)^{1/\tilde{q}'}.$$

The desired estimate follows now by summation. ∎

Proposition 9 leads to the following $L^p_w(R^n)$-norm estimate.

Theorem 10. Let ν be a weighted measure with respect to the Lebesgue measure on R^n, $d\nu(x) = w(x)dx$, and suppose $w \in A_s$, $1 < s < \infty$. Further, assume that the kernel k satisfies the condition $\tilde{M}(\tilde{q}, \tilde{\ell})$, $\tilde{\ell} > 0$ and $\tilde{q} \le \infty$, and that its associated singular integral operator T is bounded on $L^{q_0}(R^n)$, $1 \le q_0 < \infty$. Then, if $s \le p < \infty$, $q_0 \le p/s$, and $(p/s)' \le \tilde{q}$,

$$\|Tf\|_{L^p_w} \le c\|f\|_{L^p_w}, \quad f \in \hat{\mathcal{D}}_0.$$

The proof of this result follows along the lines to that of Theorem 6, but using Proposition 8 instead of Proposition 7. We leave the details to the reader.

It is also possible to make use of the basic inequality in order to obtain sharp maximal inequalities for singular integral operators. To avoid conflicts with the notation, we let

$$\tilde{M}^{\sharp,\mathcal{P}}_{d,\nu} g(x) = \sup_{x \in B} \inf_{p \in \mathcal{P}} \frac{|B|}{\nu(B)} \left(\frac{1}{|B|} \int_B |g(y) - p(y)|^d\,dy \right)^{1/d}.$$

Proposition 11. Let ν be a doubling weighted measure with respect to the Lebesgue measure on R^n, $d\nu(x) = w(x)dx$, and suppose $\nu \in D_{n\theta}$, $\theta \ge 1$. Further, let $k \in \tilde{M}(\tilde{q}, \tilde{\ell})$, $1 < \tilde{q} \le \infty$, and suppose that its associated singular integral operator T preserves $L^{q_0}(R^n)$, $1 \le q_0 < \infty$. Then, for indices d_1, d_2, so that $0 < d_1 \le q_0$ and $d_2 \ge \max(q_0, \tilde{q}')$, we have

$$\tilde{M}^{\sharp,\mathcal{P}}_{d_1,\nu}(Tf)(x) \le c\tilde{M}^{\sharp,\mathcal{P}}_{d_2,\nu} f(x), \quad f \in \hat{\mathcal{D}}_0,$$

provided that $\tilde{\ell} > n(\theta - 1)$ and $\mathcal{P} = \{$ polynomials of degree $\leq N_0\}$, with $N_0 =$ integer part of $\tilde{\ell}$.

Rough Sketch of Proof. We want to estimate

$$\frac{|B|}{\nu(B)} \left(\frac{1}{|B|} \int_B |Tf(y) - p(y)|^{d_1} \, dy \right)^{1/d_1},$$

where $B = B(0,1)$ is the unit ball. Next we consider a C^∞ splitting of k, as was done in the proof of Lemma 2, and write $k = \sum_{i=0}^{\infty} k_i$, where k_0 is supported in $2B$ and supp $k_i \subseteq \{|z| \sim 2^i\}$, $i = 1, 2, \ldots$, and $k_i \in \tilde{M}(\tilde{q}, \tilde{\ell})$ uniformly in i.

Modulo polynomials in $\mathcal{P}$ we have

$$Tf(x) = (k - k_1 - k_2 - k_3) * (\chi_{4B}(f - p_0))(x) + \sum_{i=1}^{\infty} k_i * (f - p_i)(x), \quad x \in B.$$

Here the p_j's are polynomials which will be chosen shortly, depending on the function $\tilde{M}_{d_2, \nu}^{\sharp, \mathcal{P}} f$. Using the Taylor expansion of $k_i(y - z)$ as a function of y about the origin, write $k_i(y - z) = p_{i,z}(y) + R_{i,z}(y)$, $i = 1, 2, \ldots$, where $p_{i,z}$ is a polynomial in the class $\mathcal{P}$ and $R_{i,z}(y)$, as a function of z, is supported in $\{|z| \sim 2^i\}$ and has $L^{q_0}(R^n)$-norm bounded by $c2^{-i(\tilde{\ell} + (n/\tilde{q}'))}$ when y is in B. By Hölder's inequality it then follows

$$|k_i * (f - p_i)(y)| \leq c2^{-i\tilde{\ell}} \left(2^{-in} \int_{2^{i+2}B} |f(z) - p_i(z)|^{q'} \, dz \right)^{1/q'}$$

for $y \in B$ and $i = 1, 2, \ldots$, and also

$$|k_i * (\chi_{4B}(f - p_0))(y)| \leq c \left(\int_{4B} |f(z) - p_0(z)|^{\tilde{q}'} \, dz \right)^{1/\tilde{q}'}$$

for y in B and $i = 1, 2, 3$. It is clear now that the polynomials on the right-hand side above are chosen to control, with the aid of the sharp maximal function of f, the above expressions. Also the assumption that T preserves $L^{q_0}(R^n)$ is used in estimating these expressions. By summation we obtain the desired estimate, and the proof is complete. ∎

Before we proceed we remark that Proposition 11 holds for more general functions. Indeed, if f is a measurable function and $\tilde{M}_{d_2, \nu}^{\sharp, \mathcal{P}} f$ is in $L_w^{p'}(R^n)$, $1 \leq p' \leq \infty$, we may define Tf modulo polynomials in $\mathcal{P}$ by subtracting a polynomial p_y in the variable z from the kernel $k(y - z)$ before integrating against $f(z) \, dz$. The same estimate as in Proposition 11 holds; we leave the details for the reader to verify.

Using Proposition 11 and the last remark together with the basic inequality and a duality argument, we obtain the following result.

Theorem 12. Let ν be a doubling weighted measure with respect to the Lebesgue measure on R^n, $\nu \in D_{n\theta}$, $\theta \geq 1$, and suppose $d\nu(x) = w(x)dx$, where $w \in RH_r$, $r > 1$. Further, suppose the kernel $k \in \tilde{M}(\tilde{q}, \tilde{\ell})$, $1 < \tilde{q} \leq \infty$, is such that its associated singular integral operator T preserves $L^{q_0}(R^n)$, $1 \leq q_0 < \infty$. Then, for $f \in \hat{\mathcal{D}}_0$,

$$\|Tf\|_{H^p_w} \leq c\|f\|_{H^p_w}, \quad 1 \leq p < \infty,$$

provided that $\tilde{\ell} > n(\theta - 1)$ and $pr' \leq \min(\tilde{q}, q_0')$.

Sketch of Proof. Whenever $w \in A_\infty$, from the basic inequality and Proposition 11 it follows that

$$\left| \int_{R^n} (k * f)(x)g(x)\,dx \right| \leq c \int_{R^n} M_1(F, x) M_{d_2, \nu}^{\sharp, \mathcal{P}} g(x)w(x)\,dx, \quad g \in (H^p_w(R^n))^*,$$

where $M_1(F, x)$ denotes the nontangential maximal function corresponding to an extension F of f. The condition $pr' \leq \min(\tilde{q}, q_0')$ is used to make sure that $\tilde{M}_{d_2, \nu}^{\sharp, \mathcal{P}} g$ is bounded in the $L^p_w(R^n)$-norm by the norm of g in $(H^p_w(R^n))^*$ when $d_2 = \min(\tilde{q}, q_0')$. The conclusion of the theorem follows readily from this remark; we leave the details to the reader. ∎

Also, rather than using Proposition 11 it is possible to invoke an implication between the sharp maximal function and the truncated Lusin function which essentially leads to the same duality argument as in the proof of Theorem 12; the assumptions here are that $d\nu(x) = w(x)dx$, $\nu \in D_{n\theta}$, $\theta \geq 1$, and $w \in A_\infty$. Indeed, let $f \in \hat{\mathcal{D}}_0$, and if $\psi(x) = \nabla e^{-\pi|x|^2}$, put $H(x, t) = |\psi_t * (k * f)(x)|$. Given a small positive number s, there exist positive numbers s_1, a_1 and η, depending only on s, the kernel k and w, such that for any ball B with radius h and $\eta > 0$,

$$|\{x \in B : \tilde{M}_{d_2, \nu}^{\sharp, \mathcal{P}} f(x) > a_1 a\}| < s_1|B| \quad \text{implies}$$

$$|\{x \in B : S_1^{\eta h}(H, x) > aw(x)\}| < s|B|.$$

The condition on the kernel k which is required for this implication depends only on θ and d, and the proof is left for the reader.

The methods described above have further applications. For instance, if T is a sublinear operator which maps each (q, N) atom into a sum $\sum_{i=0}^{\infty} 2^{-i\varepsilon} a_i$, say, where the a_i's are $(q, 0)$ atoms and $\varepsilon > 0$, and such that the support of the a_i's is contained in balls $B_i \subseteq 2^i B$, where $\operatorname{supp} a \subseteq B$, then T extends to a bounded mapping from $H^p_w(R^n)$ into $L^p_w(R^n)$, $0 < p < q$, provided $w \in RH_r$, $r = (q/p)'$, and $\varepsilon > \delta(w, p)$; here δ is an in Lemma 3 in Chapter VIII. These considerations apply, for instance, to pseudo-differential operators. We do not consider them in detail here since the main ideas are already apparent in the singular integral operator case.

Finally, we describe Stein's method using the Lusin and Littlewood-Paley functions to obtain pointwise estimates.

Theorem 14. Let ψ be a Schwartz function with vanishing integral and let $\phi = (\phi^{(1)}, \phi^{(2)}, \ldots, \phi^{(d)})$ be a vector-valued Schwartz function satisfying condition (1) of Chaptert V. Let $\hat{k} \in M(2, \ell)$, and for $f \in \hat{\mathcal{D}}_0$ put $F_\phi(x, t) = (\phi_t * f)(x)$ and $G_\psi(x, t) = \psi_t * (k * f)(x)$. Then,

$$S_{1,2}(G_\psi, x) \le c g_\lambda^*(F_\phi, x), \quad f \in \hat{\mathcal{D}}_0,$$

provided $\lambda \le \ell$ when ℓ is an integer, and $\lambda < \ell$ when ℓ is not an integer.

Outline of the Proof. We need only to observe that the functions $\{\eta^{(r)}\}_{r>0}$ defined by

$$\eta^{(r)}(\xi) = k(\xi/r)\psi(\xi), \quad r > 0,$$

verify the conditions of Theorem 4 in Chapter V with $A = 1$, B any positive number, and the integer $m \le \ell$. By the remarks following Theorem 8 in Chapter V we get the desired estimate when ℓ is an integer. Since the case when ℓ is not an integer was not considered in Chapter V, we give a direct proof here. We may assume that $d = 1$, that the support of ψ is contained in $\{|\xi| \sim 1\}$, and that $\phi(\xi) = 1$ on the support of ψ. Then,

$$\psi(t\xi)\hat{k}(\xi)\hat{f}(\xi) = \hat{k}(\xi)\psi(t\xi)\hat{f}(\xi)\phi(t\xi),$$

and consequently,

$$G_\psi(y, t) = (\psi_t * k) * F_\phi(\cdot, t)(y).$$

Since $\psi_t * k$ satisfies the condition $M(2, \ell)$ uniformly in $t > 0$ and is supported in $\{|\xi| \sim 1/t\}$, from the estimates in the proof of Lemma 1 it follows that

$$\int_{\{|y|\sim 2^i\}} |(\psi_t * k)(y)|^2 \, dy \le c 2^{-in}(2^i/t)^{n-2\ell}, \quad 2^i \ge t,$$

and, by Plancherel's formula, that

$$\int_{\{|y|\le t\}} |\psi_t * k(y)|^2 \, dy \le c t^{-n}, \quad t > 0.$$

Thus, $\int_{R^n} |\psi_t * k(y)|^2 (1 + (|y|/t))^{2\lambda} t^n \, dy \le c$, and we get

$$G_\psi(z, t) \le c \int_{R^n} |F_\phi(y, t)|^2 (1 + (|z - y|))^{-2\lambda} t^{-n} \, dy.$$

Substituing this estimate in the definition of $S_{1,2}(G_\psi, x)$ we obtain the desired estimate after changing the order of integration. This completes the proof. ∎

If the Schwartz function ψ satisfies in addition $|\psi(\xi)| \ge c > 0$ on the annulus $\{|\xi| \sim 1\}$ and the vector-valued function ϕ in Theorem 13 in addition satisfies $\phi(0) = 0$, then we have

$$\|k * f\|_{H_w^p} \sim \|S_{1,2}(G_\psi)\|_{L_w^p}, \quad \text{and} \quad \|f\|_{H_w^p} \sim \|S_{1,2}(F_\phi)\|_{L_w^p},$$

when $w \in A_\infty$. Using Theorem 13 together with Theorems 4 and 5 in Chapter V we get the following result.

Theorem 14. Let ν be a doubling weighted measure with respect to the Lebesgue measure on R^n, $\nu \in D_{n\theta}$, $\theta \geq 1$, and let $d\nu(x) = w(x)dx$, with $w \in A_s$, $s > 1$. Further, let $m \in M(2, \ell)$. Then, the associated multiplier operator T satisfies the estimate

$$\|Tf\|_{H_w^p} \leq c\|f\|_{H_w^p}, \quad f \in \hat{\mathcal{D}}_0,$$

provided

$$\begin{cases} 0 < p \leq 2 & \text{if } \ell > n\theta/p, \\ 2 < p < 2s & \text{if } \ell > n\frac{2s-1}{p(s-1)}(\theta - 1) + 1/2 \\ 2s \leq p < \infty & \text{if } \ell > n/2. \end{cases}$$

Sources and Remarks. Hörmander's multiplier theorem is proved in Hörmander [1960], and the use of the Lusin and Littlewood-Paley functions in this context is highlighted in Stein [1970]. Results such as the Corollary to Theorem 12 were first proved by Calderón and Torchinsky [1977]. Relations between the conditions $M(q, \ell)$ and $\tilde{M}(\tilde{q}, \tilde{\ell})$ are considered by Kurtz and Wheeden [1979].

XII

Complex Interpolation

This section is devoted to the interpolation of analytic families of operators acting on weighted Lebesgue and Hardy spaces. In this context the use of the atomic decomposition is particularly appropriate because the constructions required become quite explicit.

We remind the reader of Calderón's complex method of interpolation for Banach spaces. Let A_j, $j = 0, 1$, be complex Banach spaces continuously embedded in a Banach space A, and let $\mathcal{F} = \{f\}$ denote the class of analytic functions f defined on the strip $\Omega = \{z \in C : 0 \leq \Re(z) \leq 1\}$ with values in $A_0 + A_1$ which satisfy the following three properties:

(i) f is analytic in the interior of Ω, i.e., for $0 < \Re(z) < 1$;

(ii) f is continuous in Ω and $\|f\|_{A_0+A_1}$ is bounded in Ω;

(iii) $f(j + it) \in A_j$, and $\|f(j + it)\|_{A_j}$ is bounded, $j = 0, 1$.

Endowed with the norm

$$\|f\|_{\mathcal{F}} = \max_{j=0,1} \sup_t \|f(j + it)\|_{A_j},$$

$\mathcal{F}$ becomes a Banach space. Now, for $0 < s < 1$ let

$$A_s = \{x \in A_0 + A_1 : x = f(s), f \in \mathcal{F}\} \quad \text{and} \quad \|x\|_s = \inf_{f(s)=x} \|f\|_{\mathcal{F}}.$$

Then A_s is a Banach space which is intermediate with respect to the interpolation pair (A_0, A_1). Moreover,

Proposition 1. For $x \in A_s$ we have

$$\|x\|_{A_s} = \inf_{f(s)=x, \, f \in \mathcal{F}} \left(\sup_t \|f(it)\|_{A_0} \right)^{1-s} \left(\sup_t \|f(1 + it)\|_{A_1} \right)^s.$$

Proof. Suppose $f \in \mathcal{F}$ is such that $f(s) = x$ and for a real number λ let $g(z) = e^{\lambda(z-s)} f(z)$; thus $g \in \mathcal{F}$ and $g(s) = x$. Moreover,

$$\|g(it)\|_{A_0} = e^{-\lambda s}\|f(it)\|_{A_0}, \quad \|g(1+it)\|_{A_1} = e^{-\lambda(1-s)}\|f(1+t)\|_{A_1}.$$

Now choose λ so that $e^{\lambda} = \sup_t \|f(it)\|_{A_0} / \sup_t \|f(1+it)\|_{A_1}$. Thus,

$$\|x\|_{A_s} \le \|g\|_{\mathcal{F}} = \left(\sup_t \|f(it)\|_{A_0} \right)^{1-s} \left(\sup_t \|f(1+it)\|_{A_1} \right)^s,$$

and the inequality "$\le$" follows readily since $f \in \mathcal{F}$ is arbitrary.

Conversely, given $\varepsilon > 0$, by definition there is $f \in \mathcal{F}$ with $f(s) = x$ and $\|f\|_{\mathcal{F}} \le \|x\|_{A_s} + \varepsilon$. But,

$$\|f\|_{\mathcal{F}} \ge \left(\sup_t \|f(it)\|_{A_0} \right)^{1-s} \left(\sup_t \|f(1+it)\|_{A_1} \right)^s,$$

and the inequality "$\ge$" also holds as $\varepsilon > 0$ is arbitrary. ∎

This observation concerning the norm in A_s motivates our next result; first some notations. If θ is a real number, $0 \le \theta \le 1$, and w and v are A_∞ weights, we let

$$\theta(x) = w(x)(v(x)/w(x))^{\theta}, \quad d\theta(x) = \theta(x)dx. \tag{1}$$

Thus, informallly, the "0" measure corresponds to w, while the "1" measure corresponds to v.

Also, if $0 < p_0 \le p_1 < \infty$, and $z \in \Omega$, let $p(z)$ be defined by

$$\frac{1}{p(z)} = \frac{1-z}{p_0} + \frac{z}{p_1}. \tag{2}$$

When $0 < z = s < 1$, we simply denote $p(z) = p(s) = p$, and introduce the quantity

$$\mu(u) = \frac{up(u)}{p_1}, \quad 0 \le u \le 1. \tag{3}$$

The real numbers $\mu(u)$ and $\mu = \mu(s) = sp/p_1$ determine measures $\mu(u)$ and μ, respectively, as defined above.

With these notations we have,

Theorem 2. Suppose $f(z)$ is a function of the complex variable $z \in \Omega$ with values in $L_0^{p_0}(R^n) + L_1^{p_1}(R^n) = L_w^{p_0}(R^n) + L_v^{p_1}(R^n)$, which is uniformly continuous and bounded there, and analytic in $0 < \Re(z) < 1$. If $\sup_t \|f(j+it)\|_{L_j^{p_j}} < \infty$ for $j = 0, 1$, then $f(z) \in L_\mu^p(R^n)$ when $z = s$ and

$$\|f(s)\|_{L_\mu^p} \le \left(\sup_t \|f(it)\|_{L_0^{p_0}} \right)^{1-s} \left(\sup_t \|f(1+it)\|_{L_1^{p_1}} \right)^s.$$

Conversely, suppose f is a function in $L_\mu^p(R^n)$. Then, there is a function $f(z)$ of the complex variable z in Ω which takes values in $L_0^{p_0}(R^n) + L_1^{p_1}(R^n)$, with the following properties: $f(z)$ is uniformly continuous and bounded for $z \in \Omega$, analytic for z with $0 < \Re(z) < 1$, $f(s) = f$, and

$$\|f(u+it)\|_{L^{p(u)}_{\mu(u)}}^{p(u)} \le \|f\|_{L_\mu^p}^p, \quad u + it \in \Omega.$$

Before we prove this result we state the corresponding theorem for the weighted Hardy spaces.

Theorem 3. Assume $f(z)$ is a function of the complex variable $z \in \Omega$ which takes values in $H_0^{p_0}(R^n) + H_1^{p_1}(R^n) = H_w^{p_0}(R^n) + H_v^{p_1}(R^n)$, so that $F(x,t,z) = f(z) * \phi_t(x)$ is uniformly continuous and bounded for z in Ω and (x,t) in any compact subset of $\{(x,t): x \in R^n, t > 0\}$ and analytic for z in the interior of Ω; here ϕ is a Schwartz function with nonvanishing integral. If $f(j+it) \in H_j^{p_j}(R^n)$ and $\sup_t \|f(j+it)\|_{H_j^{p_j}} < \infty$ for $j = 0, 1$, then $f(z)$ belongs to $H_\mu^p(R^n)$ when $z = s$ and

$$\|f(s)\|_{H_\mu^p} \sim \|M_0(F(\cdot,s))\|_{L_\mu^p}$$
$$\le \left(\sup_t \|M_0(F(\cdot,it))\|_{L_0^{p_0}}\right)^{1-s} \left(\sup_t \|M_0(F(\cdot,1+it))\|_{L_1^{p_1}}\right)^s$$
$$\sim \left(\sup_t \|f(it)\|_{H_0^{p_0}}\right)^{1-s} \left(\sup_t \|f(1+it)\|_{H_1^{p_1}}\right)^s.$$

Here $M_0(F(\cdot,z))$ denotes the radial maximal function of $F(\cdot,z)$ defined with ϕ.

On the other hand, if $f \in \hat{\mathcal{D}}_0$, there is a function $f(z)$ of the variable $z \in \Omega$ such that $F(x,t,z)$ has the properties stated in the first part of the Theorem, $f(s) = f$, and

$$\|f(u+it)\|_{H^{p(u)}_{\mu(u)}} \sim \|M_0(F(\cdot,u+it))\|_{L^{p(u)}_{\mu(u)}}$$
$$\le c\|M_0(F(\cdot,s))\|_{L_\mu^p} \sim \|f\|_{H_\mu^p}, \quad u + it \in \Omega.$$

Combining Theorems 2, 3 and well-known properties of intermediate spaces we obtain the desired interpolation theorem for analytic families of operators.

Theorem 4. Let $0 < p_0, p_1 < \infty$, and suppose (A_0, A_1) is an interpolation pair of Banach spaces. Further, let $\mathcal{D}, \mathcal{D}_s, 0 \le s \le 1$, be defined by
  (i) $\mathcal{D} = A_0 \cap A_1$, and $\mathcal{D}_s = A_s, 0 \le s \le 1$; or
  (ii) $\mathcal{D} = \{f: f = \sum_j \lambda_j a_j$ where the a_j's are (∞, N) atoms with $N = N(w,v)$ appropriately chosen$\}$, and $\mathcal{D}_s = H_s^p(R^n)$; or

(iii) $\mathcal{D} = \{f : f = \sum_j \lambda_j \chi_{E_j}$ is a simple function$\}$, and $\mathcal{D}_s = L^p_\mu(R^n)$.

Also, let $0 < q_0, q_1 < \infty$, and suppose (B_0, B_1) is an interpolation pair of Banach spaces. Suppose $\mathcal{R}$ and $\mathcal{R}_s$, $0 \le s \le 1$, are defined by either

(iv) $\mathcal{R} = B_0 + B_1$, and $\mathcal{R}_s = B_s$, $0 \le s \le 1$; or

(v) $\mathcal{R} = \{f : f \in \mathcal{S}'(R^n)\}$, and $\mathcal{R}_s = H^q_{\tilde{\mu}}(R^n)$, where $1/q = (1-s)/q_0 + s/q_1$, $\tilde{\mu} = sq/q_1$, $0 \le s \le 1$; or

(vi) $\mathcal{R} = f : f$ is Lebesgue measurable$\}$, and $\mathcal{R}_s = L^q_{\tilde{\mu}}(R^n)$.

Finally, assume that $\{T_z\}$ is a family of linear operators defined on $\mathcal{D}$ with values in $\mathcal{R}$ for $z \in \Omega$. Also, suppose that, in case (iv), for each continuous linear functional ℓ on $B_0 + B_1$, $\ell(T_z f)$ is continuous and bounded for $z \in \Omega$, analytic for z in the interior of Ω and all $f \in \mathcal{D}$; in case (v), that $T_z f * \phi_t(x)$ is uniformly continuous and bounded for z in the interior of Ω and (x, t) in any compact subset of $\{(x, t) : x \in R^n, t > 0\}$ for each $f \in \mathcal{D}$; and, in case (vi), that for almost every x, $T_z f(x)$ is analytic and bounded for z in the interior of Ω and continuous in Ω.

If there are constants c_j, $j = 0, 1$, such that

$$\|T_{j+it} f\|_{\mathcal{R}_j} \le c_j \|f\|_{\mathcal{D}_j}, \quad j = 0, 1,$$

then

$$\|T_s f\|_{\mathcal{R}_s} \le c c_0^{1-s} c_1^s \|f\|_{\mathcal{D}_s}, \quad 0 < s < 1.$$

The proof of this result is left to the reader as it is essentially contained in the work of Calderón [1964]. We only remark that the constant c appears on the right-hand side due to the indetermination of the "norms" in the Hardy spaces. Before we proceed with the proof of the other results we discuss some preliminary material; first we consider a version of the Phragmen-Lindelof principle.

Lemma 5. Assume $g(z)$ is a continuous function defined on Ω, analytic in the interior of Ω, and that for some constants a, A,

$$|g(z)| \le c \exp\left(A \exp\left(a|\Im(z)|\right)\right), \quad A < \infty, \quad 0 < a < \infty.$$

Then for $0 \le s \le 1$ we have

$$\ln|g(s)| \le \int_{-\infty}^{\infty} \ln|g(it)| P_0(s, t)\, dt + \int_{-\infty}^{\infty} \ln|g(1+it)| P_1(s, t)\, dt,$$

where P_0, P_1 are the Poisson kernels for Ω.

This result is well-known and taken for granted here.

Corollary 6. Under the assumptions of Lemma 6 we have

$$|g(s)|^p \le \left(\frac{1}{1-s} \int_{-\infty}^{\infty} |g(it)|^{p_0} P_0(s, t)\, dt\right)^{1-\mu} \left(\frac{1}{s} \int_{-\infty}^{\infty} |g(1+it)|^{p_1} P_1(s, t)\, dt\right)^{\mu},$$

where $p = p(s)$ and $\mu = \mu(s)$ are as defined above.

Proof. Since

$$\int_{-\infty}^{\infty} P_0(s,t)\,dt = 1-s\,, \quad \text{and} \quad \int_{-\infty}^{\infty} P_1(s,t)\,dt = s\,,$$

from Jensen's inequality it follows that

$$|g(s)|^p \le \exp\left(\frac{p}{p_0}\int_{-\infty}^{\infty} \ln|g(it)|^{p_0} P_0(s,t)\,dt + \frac{p}{p_1}\int_{-\infty}^{\infty} \ln|g(1+it)|^{p_1} P_1(s,t)\,dt\right),$$

which in turn is dominated by

$$\exp\left(\frac{1}{1-s}\int_{-\infty}^{\infty} \ln|g(it)|^{p_0} P_0(s,t)\,dt\right)^{p(1-s)/p_0}$$

$$\times \exp\left(\frac{1}{s}\int_{-\infty}^{\infty} \ln|g(1+it)|^{p_1} P_1(s,t)\,dt\right)^{ps/p_1}$$

$$\le \left(\frac{1}{1-s}\int_{-\infty}^{\infty} |g(it)|^{p_0} P_0(s,t)\,dt\right)^{p(1-s)/p_0}$$

$$\times \left(\frac{1}{s}\int_{-\infty}^{\infty} |g(1+it)|^{p_1} P_1(s,t)\,dt\right)^{ps/p_1}.$$

Since $\mu = sp/p_1$ and $1-\mu = (1-s)p/p_0$, we are done. ∎

Proof of Theorem 2. Let $\mathcal{O}$ be a bounded subset of R^n. For almost every x in $\mathcal{O}$ we can apply Corollary 6 to the function $f(z,x)$ of z in $\mathcal{O}$ and obtain

$$\int_{\mathcal{O}} |f(s,x)|^p w(x)^{1-\mu} v(x)^{\mu}\,dx$$

$$\le \int_{\mathcal{O}} \left(\frac{1}{1-s}\int_{-\infty}^{\infty} |f(it,x)|^{p_0} P_0(s,t)\,dt\, w(x)\right)^{1-\mu}$$

$$\times \left(\frac{1}{s}\int_{-\infty}^{\infty} |f(1+it,x)|^{p_1} P_1(s,t)\,dt\, v(x)\right)^{\mu}\,dx.$$

By Hölder's inequality this last expression does not exceed

$$\left(\frac{1}{1-s}\int_{-\infty}^{\infty}\int_{\mathcal{O}} |f(it,x)|^{p_0} w(x)\,dx\, P_0(s,t)\,dt\right)^{1-\mu}$$

$$\times \left(\frac{1}{s}\int_{-\infty}^{\infty}\int_{\mathcal{O}} |f(1+it,x)|^{p_1} P_1(s,t)\,dt\right)^{\mu}$$

$$\le \left(\sup_t \|f(it)\|_{L_0^{p_0}}\right)^{(1-\mu)p_0} \left(\sup_t \|f(1+it)\|_{L_1^{p_1}}\right)^{\mu p_1} = I,$$

say. Since $(1-\mu)p_0 = p(1-s)$, $\mu p_1 = ps$, and $\mathcal{O}$ is an arbitrary bounded set, we conclude that

$$\|f(s)\|_{L^p_\mu} \leq I.$$

For the second part of the theorem, we set

$$f(z,x) = |f(x)|^{p/p(z)} \frac{f(x)}{|f(x)|} \left(\frac{w(x)}{v(x)}\right)^{(z-s)p/p_0 p_1}.$$

For almost every x, $f(x)$ is finite, and for such x's, $f(z,x)$ is a continuous bounded function of z in Ω, analytic in the interior of Ω. Also, since $(u-s)p/p_0 p_1 = (\mu(u)-\mu)/p(u)$,

$$\|f(u+it)\|_{L^{p(u)}_{\mu(u)}} \leq \|f\|^p_{L^p_\mu}, \quad u+it \in \Omega.$$

Moreover, since for fixed $u+it \in \Omega$ we have

$$|f(u+it,x)| = |f(0,x)|^{1-u}|f(1,x)|^u,$$

for each x we have

$$|f(u+it,x)| \leq \max\left(|f(0,x)|, |f(1,x)|\right).$$

Let $\mathcal{E} = \{x: |f(u+it,x)| \leq |f(0,x)|\}$. We can then write

$$\begin{aligned} f(u+it,x) &= \chi_\mathcal{E}(x)f(u+it,x) + (1-\chi_\mathcal{E}(x))f(u+it,x) \\ &= f_0 + f_1, \end{aligned}$$

say. Clearly $f_j \in L^{p_j}_j(R^n)$, and

$$\|f_j\|^{p_j}_{L^{p_j}_j} \leq \|f(j)\|^{p_j}_{L^{p_j}_j} = \|f\|^p_{L^p_\mu}, \quad j=0,1.$$

This completes the proof. ∎

Proof of Theorem 3. For the first half of the theorem we let ϕ be a Schwartz function with integral 1, and $t(x)$ an arbitrary, positive, measurable function of x. We now apply the first part of Theorem 3 to the function

$$F(z,x) = f(z) * \phi_{t(x)}(x).$$

Since for fixed $z \in \Omega$, $f(z) = f_0 + f_1$ with $f_j \in H^{p_j}_j(R^n)$, $j=0,1$, and

$$|f_j * \phi_{t(x)}(x)| \leq M_0(f_j * \phi_t, x), \quad j=0,1,$$

it follows that

$$F(z,x) = f_0 * \phi_{t(x)}(x) + f_1 * \phi_{t(x)}(x) = F_0(x) + F_1(x),$$

say, with

$$\|F_j\|_{L_j^{p_j}} \le c\|M_0(f_j * \phi_t)\|_{L_j^{p_j}} \le c\|f_j\|_{H_j^{p_j}}, \quad j = 0,1.$$

Furthermore, observe that

$$\|F(j+it)\|_{L_j^{p_j}} \le c\|M_0(f(j+it) * \phi_t)\|_{L_j^{p_j}} \le c\|f(j+it)\|_{H_j^{p_j}}, \quad j = 0,1.$$

For all x, $F(z,x)$ is a continuous bounded function of z in Ω which is analytic in the interior of Ω. Thus, from Theorem 3 we conclude that

$$\|F(s)\|_{L_\mu^p} \le \left(\sup_t \|F(it)\|_{L_0^{p_0}}\right)^{1-s} \left(\sup_t \|F(1+it)\|_{L_1^{p_1}}\right)^s.$$

Taking the supremum over all measurable functions $t(x)$, since $\|M_0(f(s))\|_{L_\mu^p} = \sup_{t(x)} \|f(s)\|_{L_\mu^p}$, it follows that

$$\|f(s)\|_{H_\mu^p} \sim \|M_0(f(s))\|_{L_\mu^p}$$

$$\le c \left(\sup_t \|M_0(f(it))\|_{L_0^{p_0}}\right)^{1-s} \left(\sup_t \|M_0(f(1+it))\|_{L_1^{p_1}}\right)^s$$

$$\sim c \left(\sup_t \|f(it)\|_{H_0^{p_0}}\right)^{1-s} \left(\sup_t \|f(1+it)\|_{H_1^{p_1}}\right)^s.$$

The proof of the first half of Theorem 3 is thus complete. As for the proof of the second half, we need some preliminary results.

Lemma 7. Suppose w and v are A_∞ weights, and $0 \le \delta \le 1$. Then $w^{1-\delta}v^\delta$ is an A_∞ weight and

$$\frac{1}{|B|} \int_B w(x)^{1-\delta} v(x)^\delta \, dx \sim \left(\frac{1}{|B|} \int_B w(x)\, dx\right)^{1-\delta} \left(\frac{1}{|B|} \int_B v(x)\, dx\right)^\delta$$

for all balls B with the constant in $\sim$ independent of B.

Proof. If δ is equal to 0 or 1 there is nothing to prove. Since w,v are in A_q for some $q < \infty$, it is an exercise using Hölder's inequality to show that the right-hand side in the conclusion of the lemma dominates the left-hand side.

As for the inequality in the other direction, we set $\varepsilon = 1/(1+\delta(q-1))$, where $v \in A_q$. Also, since $w \in A_\infty$, we have

$$\left(\frac{1}{|B|}\int_B w(x)\,dx\right)^{\varepsilon(1-\delta)} \sim \left(\frac{1}{|B|}\int_B w(x)^{\varepsilon(1-\delta)}\,dx\right).$$

Writing $w(x)^{\varepsilon(1-\delta)} = (w(x)^{\varepsilon(1-\delta)}v(x)^{\varepsilon\delta})v(x)^{-\varepsilon\delta}$, from Hölder's inequality we get

$$\left(\frac{1}{|B|}\int_B w(x)\,dx\right)^{\varepsilon(1-\delta)}$$
$$\leq c\left(\frac{1}{|B|}\int_B w(x)^{1-\delta}v(x)^\delta\,dx\right)^\varepsilon \left(\frac{1}{|B|}\int_B v(x)^{-\varepsilon\delta/(1-\varepsilon)}\,dx\right)^{1-\varepsilon}.$$

Since $(1-\varepsilon)/\varepsilon = \delta(q-1)$ and $-\varepsilon\delta/(1-\varepsilon) = -1/(q-1)$, it follows that

$$\left(\frac{1}{|B|}\int_B w(x)\,dx\right)^{1-\delta}$$
$$\leq c\left(\frac{1}{|B|}\int_B w(x)^{1-\delta}v(x)^\delta\,dx\right)\left(\frac{1}{|B|}\int_B v(x)^{-1/(q-1)}\,dx\right)^{\delta(q-1)}$$
$$\leq c\left(\frac{1}{|B|}\int_B w(x)^{1-\delta}v(x)^\delta\,dx\right)\left(\frac{1}{|B|}\int_B v(x)\,dx\right)^{-\delta},$$

where in the last inequality we used the fact that $v \in A_q$. ∎

Lemma 8. Let ν be a weighted measure with respect to the Lebesgue measure on R^n, $d\nu(x) = w(x)dx$, $w \in A_\infty$, and suppose $0 < p < \infty$. Then

$$\left\|\sum_k \lambda_k(\nu(B_k)/|B_k|)^{1/p}\chi_{B_k}\right\|_{L^p} \sim \left\|\sum_k \lambda_k\chi_{B_k}\right\|_{L^p_w},$$

for all collections $\{B_k\}$ of balls and all choices of positive real numbers λ_k. The constant in the relation $\sim$ depends only on p and w.

Proof. First recall the following (minor variant of a) result in Chapter VIII: Let $0 < p < \infty$ and $0 < s \leq 1$. Then

$$\left\|\sum_k \lambda_k\chi_{E_k}\right\|_{L^p_w} \sim \left\|\sum_k \lambda_k\chi_{B_k}\right\|_{L^p_w},$$

for all collections $\{B_k\}$ of balls and all collections $\{E_k\}$ of measurable subsets $E_k \subseteq B_k$ of the B_k's such that $\nu(E_k) \geq \eta\nu(B_k)$, $\eta > 0$. The constant in the relation $\sim$ depends only on w, p and η.

With this remark out of the way, for $r > 0$ put

$$E_k = \{x \in B_k : \nu(B_k)/r|B_k| \le w(x) \le r\nu(B_k)/|B_k|\}.$$

Since $w \in A_\infty$, we get that $|E_k| \ge |B_k|/2$ and $\nu(B_k) \ge |B_k|/2$ if r is chosen large enough, the choice depending on w. By the preceding remark we get

$$\left\|\sum_k \lambda_k (\nu(B_k)/|B_k|)^{1/p} \chi_{B_k}\right\|_{L^p} \sim \left\|\sum_k \lambda_k (\nu(B_k)/|B_k|)^{1/p} \chi_{E_k}\right\|_{L^p}$$

$$\sim \left\|\sum_k \lambda_k \chi_{E_k} w^{1/p}\right\|_{L^p} = \left\|\sum_k \lambda_k \chi_{E_k}\right\|_{L^p_w}$$

$$\sim \left\|\sum_k \lambda_k \chi_{B_k}\right\|_{L^p_w}.$$

This completes the proof. ∎

Proof of the second half of Theorem 3. Let $f = \sum_k \lambda_k a_k$, supp $a_k \subseteq B_k$, denote the atomic decomposition of f into (∞, N) atoms, with an appropriately large N; we can also use (q, N) atoms for q and N sufficiently large. This decomposition may be assumed to have the further property that

$$\left|\sum_k \lambda_k^r \chi_{B_k}(x)\right| \le c_r f^*(x)^r, \quad \text{all} \quad r > 0. \tag{4}$$

To define $f(z)$ we put

$$\lambda_k(z) = \lambda_k^{p/p(z)} \left(\frac{\nu(B_k)}{\tilde{\nu}(B_k)}\right)^{(z-s)p/p_0 p_1},$$

where $d\tilde{\nu}(x) = v(x)dx$, and set

$$f(z) = \sum_k \lambda_k(z) a_k.$$

We begin by checking that $f(z)$ has the desired properties. Clearly $f(s) = f$ and for $u + it \in \Omega$, by Lemma 8 we have

$$\|f(u+it)\|_{H^{p(u)}_{\mu(u)}}^{p(u)} \le c \left\|\sum_k |\lambda_k(u+it)| \chi_{B_k}\right\|_{L^{p(u)}_{\mu(u)}}^{p(u)}$$

which in turn is dominated by

$$c\left\|\sum_k |\lambda_k(u+it)| \left(\frac{1}{|B_k|}\int_{B_k} w(x)^{1-\mu(u)} v(x)^{\mu(u)}\,dx\right)^{1/p(u)} \chi_{B_k}\right\|_{L^{p(u)}}^{p(u)}$$

$$\le c\left\|\sum_k \lambda_k^{p/p(u)} (\nu(B_k)/\tilde{\nu}(B_k))^{(u-s)p/p_0 p_1}\right.$$

$$\left.\times \left(\frac{1}{|B_k|}\int_{B_k} w(x)^{1-\mu(u)} v(x)^{\mu(u)}\,dx\right)^{1/p(u)} \chi_{B_k}\right\|_{L^{p(u)}}^{p(u)}.$$

Since $(u - s)p/p_0 p_1 = (\mu(u) - \mu)/p(u)$, by Lemma 7 we get

$$\left(\frac{\nu(B_k)}{\tilde{\nu}(B_k)}\right)^{(u-s)p/p_0 p_1} \left(\frac{1}{|B_k|} \int_{B_k} w(x)^{1-\mu(u)} v(x)^{\mu(u)}\, dx\right)^{1/p(u)}$$

$$\leq c\nu(B_k)^{1-\mu/p(u)} \tilde{\nu}(B_k)^{\mu/p(u)} |B_k|^{1/p(u)}$$

$$\leq c\left(\frac{1}{|B_k|} \int_{B_k} w(x)^{1-\mu} v(x)^{\mu}\, dx\right)^{1/p(u)}.$$

Whence,

$$\|f(u+it)\|_{H^{p(u)}_{\mu(u)}}^{p(u)} \leq c\left\|\sum_k \lambda_k^{p/p(u)} \left(\frac{1}{|B_k|} \int_{B_k} w(x)^{1-\mu} v(x)^{\mu}\, dx\right)^{1/p(u)} \chi_{B_k}\right\|_{L^{p(u)}}^{p(u)}$$

$$\leq c\left\|\sum_k \lambda_k^{p/p(u)} \chi_{B_k}\right\|_{L^{p(u)}_\mu}^{p(u)}.$$

Thus, by (4) with $r = p/p(u)$ there, we get

$$\|f(u+it)\|_{H^{p(u)}_{\mu(u)}}^{p(u)} \leq c\|(f^*)^{p/p(u)}\|_{L^{p(u)}_{\mu(u)}}^{p(u)} \leq c\|f^*\|_{L^p_\mu}^p \leq c\|f\|_{H^p_\mu}^p.$$

In the case $0 < p_0 \leq p_1 \leq 1$, the above estimates can be obtained in a somewhat simpler fashion, without making use of Lemmas 8 and 9, which are fairly complicated when $0 < p < 1$. Indeed, we have

$$\|f(u+it)\|_{H^{p(u)}_{\mu(u)}}^{p(u)} \leq c\sum_k |\lambda_k(u+it)|^{p(u)} \nu(B_k)$$

$$= \sum_k \lambda_k^p (\tilde{\nu}(B_k)/\nu(B_k))^u \nu(B_k),$$

which, by Lemma 7, is less than or equal to

$$c\sum_k \lambda_k^p \left(\int_{B_k} w(x)^{1-\mu} v(x)^\mu\, dx\right) \leq c\|f\|_{H^p}^p.$$

Next we show that $f(z) \in H^{p_0}_w(R^n) + H^{p_1}_v(R^n)$ for z in Ω. Fix $u + it \in \Omega$ and observe that

$$|\lambda_k(u+it)| = \lambda_k(0)^{1-u} \lambda_k(1)^u.$$

Thus, for any k, at least one of the inequalities

$$|\lambda_k(u+it)| \leq \lambda_k(j), \quad j = 0, 1,$$

holds. Setting $f_0 = \sum_k \lambda_k(u+it) a_k$, where the sum is taken over those k's for which $|\lambda_k(u+it)| \leq \lambda_k(0)$, and $f_1 = f - f_0$, we get that $f_j \in H^{p_j}_j(R^n)$, $j = 0, 1$, and

$$\|f_j\|_{H^{p_j}_j} \leq c\left\|\sum_k \lambda_k(j) \chi_{B_k}\right\|_{L^{p_j}_j} \leq c\|f\|_{H^p_\mu}^{p/p_j}, \quad j = 0, 1.$$

It remains to show that $F(z, x, t)$ is uniformly continuous and bounded when (x, t) lies in any compact subset K of $\{(x, t) : t > 0\}$, and analytic for z in the interior of Ω. Given $\varepsilon > 0$, let $f_N = \sum_{k=1}^{N} \lambda_k a_k$ be a finite sum of atoms such that

$$\|f - f_N\|_{H_\mu^p} \leq c \left\| \sum_{k=N+1}^{\infty} \lambda_k \chi_{B_k} \right\|_{L_\mu^p} < \varepsilon .$$

Now, it is readily checked that

$$F_N(z, x, t) = f_N(z) * \phi_t(x) = \sum_{k=1}^{N} \lambda_k(z)(a_k * \phi_t)(x)$$

satisfies the properties we stated for $F(z, x, t)$. Since for $(x, t) \in K$ we have

$$|g * \phi_t(x)| \leq c_{K,\phi} \|g\|_{H_j^{p_j}} , \quad j = 0, 1 ,$$

it follows that

$$|F(z, x, t) - F_N(z, x, t)| \leq c(\varepsilon^{p/p_0} + \varepsilon^{p/p_1}) .$$

Here we used the fact that $f(z) - f_N(z)$ belongs to $H_0^{p_0}(R^n) + H_1^{p_1}(R^n)$, with norm less than or equal to $c(\varepsilon^{p/p_0} + \varepsilon^{p/p_1})$, which follows directly from the above estimates.

By a limiting argument using Morera's theorem we conclude that $F(z, x, t)$ has the desired properties. The proof of Theorem 4 is thus complete. ∎

We conclude this chapter with three remarks. First, the proof of the second half of Theorem 4 relies on the fact that our "nice" function f can be decomposed into a sum of atoms that satisfy the estimate given in (4). When dealing with an arbitrary function $f \in H_\mu^p(R^n)$, we can approximate it by a sum $\sum_{j=1}^{\infty} f_j$ of "nice" functions f_j where $\|f - \sum_{j=1}^{N} f_j\|_{H_\mu^p} \leq 2^{-N}$ and $\|f_j\|_{H_\mu^p} \leq c2^{-j} \|f\|_{H_\mu^p}$, decompose each function f_j, and then write $f = \sum_k \lambda_k a_k$ as a sum of atoms such that

$$\sum_k \lambda_k^r \chi_{B_k}(x) \leq c \sum_{j=1}^{\infty} f_j^*(x)^r , \quad 0 < r < \infty .$$

It then readily follows that for such r's,

$$\left\| \left(\sum_{j=1}^{\infty} (f_j^*)^r \right)^{1/r} \right\|_{L_\mu^p} \leq c_{r,p} \|f\|_{H_\mu^p} .$$

Next note that it is possible to redefine the $\lambda_k(z)$'s in the second part of Theorem 4 so that an estimate for $\sum_k \lambda_k^r \chi_{B_k}$, $0 < r < 1$, can be carried out in terms of the "atomic" norm associated to $\sum_k \lambda_k a_k$. Indeed, let

$$E_k \{x : \sum_k \lambda_k \chi_{5B_k}(x) > 2^j\}, \quad \text{and} \quad I_j = \{k : B_k \subseteq E_j\} .$$

Then,

$$\bigcup_{j=-\infty}^{\infty} I_j = \{\text{all } k\text{'s that appear in the sum}\},$$

and

$$\bigcap_{j=1}^{\infty} I_j = \emptyset, \quad \text{provided } \sum_k \lambda_k \chi_{5B_k}(x) < \infty \text{ a.e.}$$

Let now

$$\lambda_k(z) = 2^{j(p/p(z)-1)} \lambda_k \left(\frac{\nu(B_k)}{\tilde{\nu}(B_k)}\right)^{(z-s)p/p_1 p_0}, \quad \text{whenever } k \in I_j \setminus I_{j-1}.$$

It is not hard to show that

$$\sum_{j=-\infty}^{\infty} \sum_{k \in I_j \setminus I_{j-1}} 2^{j(r-1)} \lambda_k \chi_{B_k}(x) \leq \sum_k \lambda_k \chi_{5B_k}(x),$$

and

$$\left(\sum_k \lambda_k \chi_{B_k}(x)\right)^r \leq \sum_{j=-\infty}^{\infty} \sum_{k \in I_j \setminus I_{j-1}} 2^{j(r-1)} \lambda_k \chi_{5B_k}(x).$$

The argument used in the proof of Theorem 4 together with these pointwise estimates allow us to show that

$$\left\|\sum_k \lambda_k(u+it)\chi_{B_k}\right\|_{L_{\mu(u)}^{p(u)}}^{p(u)} \sim \left\|\sum_k \lambda_k \chi_{B_k}\right\|_{L_\mu^p}^p.$$

Finally, because when $1 \leq p < \infty$, $L_w^p(R^n)$ and $H_w^p(R^n)$ are Banach spaces, we may verify the analiticity of functions $f(z)$ taking values in $L_w^{p_0}(R^n) + L_v^{p_1}(R^n)$, and similarly for $H_w^{p_0}(R^n) + H_v^{p_1}(R^n)$, $1 \leq p_0, p_1 < \infty$, by checking that $\langle \ell, f(z) \rangle$ is analytic for every bounded linear functional ℓ. Similar remarks apply to continuity and boundedness.

Sources and Remarks. The interpolation of operators with change of measure was first established, in the context of the Lebesgue spaces, by E. Stein and G. Weiss [1958]. These authors also considered complex interpolation of operators acting in Hardy spaces of the disk, E. Stein and G. Weiss [1957]; for the Euclidean space this was done by A. Calderón and A. Torchinsky [1977]. The abstract theory of complex interpolation was developed by A. Calderón [1964].

Bibliography

Aguilera, N., and Segovia, C.
[1977] Weighted norm inequalities relating the g_λ^* and the area functions, *Studia Math.* **61**, 293-303.

Berman, S.
[1975] Characterization of bounded mean oscillation, *Proc. Amer. Math. Soc.* **5**, 117-122.

Burkholder, D., and Gundy, R.
[1972] Distribution function inequalities for the area integral, *Studia Math.* **44**, 527-544.

Burkholder, D., Gundy, R., and Silverstein, M.
[1971] A maximal function characterization of the class H^p, *Trans. Amer. Math. Soc.* **157**, 137-153.

Calderón, A.
[1964] Intermediate spaces and interpolation, *Studia Math.* **24**, 113-190.
[1976] Inequalities for the maximal function relative to a metric, *Studia Math.* **49**, 297-306.

Calderón, A., and Torchinsky, A.
[1975] Parabolic maximal functions associated with a distribution, *Advances in Math.* **16**, 1-64.
[1977] Parabolic maximal functions associated with a distribution, II, *Advances in Math.* **24**, 101-171.

Calderón, A., and Zygmund, A.
[1952] On the existence of certain singular integrals, *Acta Math.* **88**, 85-139.

Coifman, R.
[1974] A real variable characterization of H^p, *Studia Math.* **51**, 269-274.

Coifman, R., and Fefferman, C.
[1974] Weighted norm inequalities for maximal functions and singular integrals, *Studia Math* **51**, 241-250.

Coifman, R., and Weiss, G.
[1977] Extensions of Hardy spaces and their use in Analysis, *Bull. Amer. Math. Soc.* **83**, 569-645.

Duren, P., Romberg, B., and Shields, A.

[1969] Linear functionals on H^p spaces with $0 < p < 1$, *J. Reine Angew. Math.* **238**, 32-60.

Fefferman, C.

[1970] Inequalities for strongly singular integrals, *Acta Math.* **124**, 9-36.

Fefferman, C., and Stein, E.

[1971] H^p spaces of several real variables, *Acta Math.* **129**, 137-193.

García-Cuerva, J.

[1979] Weighted H^p spaces, *Dissertationes Math.* **162**.

García-Cuerva, J., and Rubio de Francia, J.

[1985] "Weighted Norm Inequalities and Related Topics," North Holland, Amsterdam.

Gehring, F.

[1973] The L^p-integrability of the partial derivatives of a quasiconformal mapping, *Acta Math.* **130**, 265-277.

Goldberg, D.

[1979] A local version of real Hardy spaces, *Duke Math. J.* **46**, 27-42.

Gundy, R., and Wheeden, R.

[1974] Weighted integral inequalities for the nontangential maximal function, Lusin area integral, and Walsh-Paley series, *Studia Math.* **49**, 107-124.

Hörmander, L.

[1960] Estimates for translation invariant operators in L^p spaces, *Acta Math.* **104**, 93-140.

Hunt, R., Muckenhoupt, B., and Wheeden, R.

[1973] Norm weighted inequalities for the conjugate function and Hilbert transforms, *Trans. Amer. Math. Soc.* **176**, 227-251.

John, F.

[1964] Quasi-isometric mappings, *Sem. Instituto Nazionali di Alta Matematica, 1962-1963*, 462-473.

John, F., and, Nirenberg, L.

[1961] On functions of bounded mean oscillation, *Comm. Pure Appl. Math.* **14**, 415-426.

Kurtz, D., and Wheeden, R.

[1979] Results on weighted norm inequalities for multipliers, *Trans. Amer. Math. Soc.* **255**, 343-362.

Latter, R.

[1977] A characterization of $H^p(R^n)$ in terms of atoms, *Studia Math.* **62**, 92-101.

Lotkowski, E., and Wheeden, R.

[1976] The equivalence of various Lipschitz conditions on the weighted mean oscillation of a function, *Proc. Amer. Math. Soc.* **61**, 323-328.

Marshall, B.

[1980] Tempered nontangential boundedness, *Studia Math.* **67**, 241-277.

Muckenhoupt, B.

[1972] Weighted norm inequalities for the Hardy maximal function, *Trans. Amer. Math. Soc.* **165**, 207-226.

[1974] The equivalence of two conditions for weight functions, *Studia Math.* **49**, 101-106.

Muckenhoupt, B., and Wheeden, R.

[1974] Norm inequalities for the Littlewood-Paley g_λ^* function, *Trans. Amer. Math. Soc.* **191**, 95-111.

[1978] On the dual of H^1 of the half-space, *Studia Math.* **63**, 57-79.

Rivière, N.

[1971] Singular integrals and multiplier operators, *Arkiv Mat.* **9**, 243-278.

Stein, E.

[1970] "Singular Integrals and Differentiability Properties of Functions," Princeton Univ. Press, Princeton, New Jersey.

Stein, E., and Weiss, G.

[1957] On the interpolation of analytic families of operators acting on H^p spaces, *Tôhoku Math. J.* **9**, 318-339.

[1958] Interpolation of operators with change of measure, *Trans. Amer. Math. Soc.* **87**, 159-172.

Strömberg, J.-O.

[1979a] Bounded mean oscillation with Orlicz norms and duality of Hardy spaces, *Indiana Math. J.* **28**, 511-544.

[1979b] Nonequivalence between two kinds of conditions on weight functions, *Proc. Symp. Pure Math* Vol. 35, Part I, *Amer. Math. Soc.*, Providence, R.I., 141-148.

Strömberg, J.-O. and Torchinsky, A.

[1980] Weights, sharp maximal functions and Hardy spaces, *Bull. Amer. Math. Soc.* **3**, 1053-1056.

Torchinsky, A.

[1979] Weighted norm inequalities for the Littlewood-Paley function g_λ^*, *Proc. Symp. Pure Math* Vol 35, Part I, *Amer. Math. Soc.*, Providence, R.I., 125-131.

[1986] "Real-Variable Methods in Harmonic Analysis," Academic Press, New York.

Wheeden, R.

[1976] A boundary value characterization of weighted H^1, *L'Enseignement Math.* **24**, 121-134.

Zygmund, A.

[1959] "Trigonometric Series," Second Edition, Volumes I and II, Cambridge Univ. Press, New York.

Index

LECTURE NOTES IN MATHEMATICS
Edited by A. Dold and B. Eckmann

Some general remarks on the publication of monographs and seminars

In what follows all references to monographs, are applicable also to multiauthorship volumes such as seminar notes.

§1. Lecture Notes aim to report new developments - quickly, informally, and at a high level. Monograph manuscripts should be reasonably self-contained and rounded off. Thus they may, and often will, present not only results of the author but also related work by other people. Furthermore, the manuscripts should provide sufficient motivation, examples and applications. This clearly distinguishes Lecture Notes manuscripts from journal articles which normally are very concise. Articles intended for a journal but too long to be accepted by most journals, usually do not have this "lecture notes" character. For similar reasons it is unusual for Ph.D. theses to be accepted for the Lecture Notes series.

Experience has shown that English language manuscripts achieve a much wider distribution.

§2. Manuscripts or plans for Lecture Notes volumes should be submitted either to one of the series editors or to Springer-Verlag, Heidelberg. These proposals are then refereed. A final decision concerning publication can only be made on the basis of the complete manuscripts, but a preliminary decision can usually be based on partial information: a fairly detailed outline describing the planned contents of each chapter, and an indication of the estimated length, a bibliography, and one or two sample chapters - or a first draft of the manuscript. The editors will try to make the preliminary decision as definite as they can on the basis of the available information.

§3. Lecture Notes are printed by photo-offset from typed copy delivered in camera-ready form by the authors. Springer-Verlag provides technical instructions for the preparation of manuscripts, and will also, on request, supply special staionery on which the prescribed typing area is outlined. Careful preparation of the manuscripts will help keep production time short and ensure satisfactory appearance of the finished book. Running titles are not required; if however they are considered necessary, they should be uniform in appearance. We generally advise authors not to start having their final manuscripts specially tpyed beforehand. For professionally typed manuscripts, prepared on the special stationery according to our instructions, Springer-Verlag will, if necessary, contribute towards the typing costs at a fixed rate.

The actual production of a Lecture Notes volume takes 6-8 weeks.

.../...

§4. Final manuscripts should contain at least 100 pages of mathematical text and should include
- a table of contents
- an informative introduction, perhaps with some historical remarks. It should be accessible to a reader not particularly familiar with the topic treated.
- a subject index; this is almost always genuinely helpful for the reader.

§5. Authors receive a total of 50 free copies of their volume, but no royalties. They are entitled to purchase further copies of their book for their personal use at a discount of 33.3 %, other Springer mathematics books at a discount of 20 % directly from Springer-Verlag.

Commitment to publish is made by letter of intent rather than by signing a formal contract. Springer-Verlag secures the copyright for each volume.

LECTURE NOTES

ESSENTIALS FOR THE PREPARATION
OF CAMERA-READY MANUSCRIPTS

Springer
Springer-Verlag
Berlin Heidelberg New York
London Paris Tokyo Hong Kong

The preparation of manuscripts which are to be reproduced by photo-offset require special care. Manuscripts which are submitted in technically unsuitable form will be returned to the author for retyping. There is normally no possibility of carrying out further corrections after a manuscript is given to production. Hence it is crucial that the following instructions be adhered to closely. If in doubt, please send us 1 - 2 sample pages for examination.

General. The characters must be uniformly black both within a single character and down the page. Original manuscripts are required: photocopies are acceptable only if they are sharp and without smudges.

On request, Springer-Verlag will supply special paper with the text area outlined. The standard TEXT AREA (OUTPUT SIZE if you are using a 14 point font) is 18 x 26.5 cm (7.5 x 11 inches). This will be scale-reduced to 75% in the printing process. If you are using computer typesetting, please see also the following page.

Make sure the TEXT AREA IS COMPLETELY FILLED. Set the margins so that they precisely match the outline and type right from the top to the bottom line. (Note that the page number will lie outside this area). Lines of text should not end more than three spaces inside or outside the right margin (see example on page 4).

Type on one side of the paper only.

Spacing and Headings (Monographs). Use ONE-AND-A-HALF line spacing in the text. Please leave sufficient space for the title to stand out clearly and do NOT use a new page for the beginning of subdivisons of chapters. Leave THREE LINES blank above and TWO below headings of such subdivisions.

Spacing and Headings (Proceedings). Use ONE-AND-A-HALF line spacing in the text. Do not use a new page for the beginning of subdivisons of a single paper. Leave THREE LINES blank above and TWO below headings of such subdivisions. Make sure headings of equal importance are in the same form.

The first page of each contribution should be prepared in the same way. The title should stand out clearly. We therefore recommend that the editor prepare a sample page and pass it on to the authors together with these instructions. Please take the following as an example. Begin heading 2 cm below upper edge of text area.

MATHEMATICAL STRUCTURE IN QUANTUM FIELD THEORY

John E. Robert
Mathematisches Institut, Universität Heidelberg
Im Neuenheimer Feld 288, D-6900 Heidelberg

Please leave THREE LINES blank below heading and address of the author, then continue with the actual text on the same page.

Footnotes. These should preferable be avoided. If necessary, type them in SINGLE LINE SPACING to finish exactly on the outline, and separate them from the preceding main text by a line.

Symbols. Anything which cannot be typed may be entered by hand in BLACK AND ONLY BLACK ink. (A fine-tipped rapidograph is suitable for this purpose; a good black ball-point will do, but a pencil will not). Do not draw straight lines by hand without a ruler (not even in fractions).

Literature References. These should be placed at the end of each paper or chapter, or at the end of the work, as desired. Type them with single line spacing and start each reference on a new line. Follow "Zentralblatt für Mathematik"/"Mathematical Reviews" for abbreviated titles of mathematical journals and "Bibliographic Guide for Editors and Authors (BGEA)" for chemical, biological, and physics journals. Please ensure that all references are COMPLETE and ACCURATE.

IMPORTANT

Pagination. For typescript, <u>number pages in the upper right-hand corner in LIGHT BLUE OR GREEN PENCIL ONLY</u>. The printers will insert the final page numbers. For computer type, you may insert page numbers (1 cm above outer edge of text area).

It is safer to number pages AFTER the text has been typed and corrected. Page 1 (Arabic) should be THE FIRST PAGE OF THE ACTUAL TEXT. The Roman pagination (table of contents, preface, abstract, acknowledgements, brief introductions, etc.) will be done by Springer-Verlag.

If including running heads, these should be aligned with the inside edge of the text area while the page number is aligned with the outside edge noting that <u>right</u>-hand pages are <u>odd</u>-numbered. Running heads and page numbers appear on the same line. Normally, the running head on the left-hand page is the chapter heading and that on the right-hand page is the section heading. Running heads should <u>not</u> be included in proceedings contributions unless this is being done consistently by all authors.

Corrections. When corrections have to be made, cut the new text to fit and paste it over the old. White correction fluid may also be used.

Never make corrections or insertions in the text by hand.

If the typescript has to be marked for any reason, e.g. for provisional page numbers or to mark corrections for the typist, this can be done VERY FAINTLY with BLUE or GREEN PENCIL but NO OTHER COLOR: these colors do not appear after reproduction.

COMPUTER-TYPESETTING. Further, to the above instructions, please note with respect to your printout that
- the characters should be sharp and sufficiently black;
- it is not strictly necessary to use Springer's special typing paper. Any white paper of reasonable quality is acceptable.

If you are using a significantly different font size, you should modify the output size correspondingly, keeping length to breadth ratio 1 : 0.68, so that scaling down to 10 point font size, yields a text area of 13.5 x 20 cm (5 3/8 x 8 in), e.g.

Differential equations.: use output size 13.5 x 20 cm.

Differential equations.: use output size 16 x 23.5 cm.

Differential equations.: use output size 18 x 26.5 cm.

Interline spacing: 5.5 mm base-to-base for 14 point characters (standard format of 18 x 26.5 cm).
If in any doubt, please send us 1 - 2 sample pages for examination. We will be glad to give advice.